ENZYMES AND FOOD

ENZYMES AND FOOD

Shahina Naz

OXFORD
UNIVERSITY PRESS

OXFORD

UNIVERSITY PRESS

Great Clarendon Street, Oxford OX2 6DP

Oxford University Press is a department of the University of Oxford.
It furthers the University's objective of excellence in research, scholarship,
and education by publishing worldwide in

Oxford New York

Auckland Bangkok Buenos Aires Cape Town Chennai
Dar es Salaam Delhi Hong Kong Istanbul Karachi Kolkata
Kuala Lumpur Madrid Melbourne Mexico City Mumbai Nairobi
São Paulo Shanghai Singapore Taipei Tokyo Toronto

and an associated company in Berlin

Oxford is a registered trade mark of Oxford University Press
in the UK and in certain other countries

ISBN 0 19 579773 6

Typeset in Times
Printed in Pakistan by
Mas Printers, Karachi.
Published by
Ameena Saiyid, Oxford University Press
5-Bangalore Town, Sharae Faisal
PO Box 13033, Karachi-75350, Pakistan.

To

Late Dr S.A. Zubairi
Professor, Department of Chemistry
University of Karachi

Contents

SECTION A: ENZYMES IN GENERAL

PART I : ENZYME AS PROTEIN

PART II: ENZYME AS CATALYST

PART III: INVESTIGATION AND PURIFICATION OF ENZYMES

SECTION B: ENZYMES IN FOOD

PART I: OXIDOREDUCTASES

Foreword

Substantial literature is available on sources, structure, properties, laboratory production, purification, and theoretical behaviour of enzymes. However little attention has been paid towards writing and organizing a comprehensive script that could provide information ranging from basic science of enzymes to their importance and practical applications in food industries, thus giving an impact of the slow merging of general enzymology into food technology. *Enzymes and Food* fills this void. This text is a unique collection of information covering basic definition, nomenclature, structure, properties, isolation and purification and specific applications of various enzymes in food industries. As such it will be of value to a wide variety of workers in industry as well as academia.

All chapters have been updated and the literature is covered until late 1990s. Even the latest applications of enzymes in food have been included.

I hope that this book will be useful not only for food scientists but also for students, teachers and researchers in the fields in which enzymes are important.

Dr Viqar Uddin Ahmad
Professor, HEJ Research Institute of Chemistry
University of Karachi

Preface

Enzymology is such a vast and rapidly growing science that writing a textbook covering all the various theoretical and applied aspects of the subject is a big undertaking. Though this book is intended mainly for students taking degree courses in Food Science and Technology, large portions may be of value to those who may be approaching the subject of enzymology for the first time.

Part I of Section 'A' comprises general introduction to enzymes; Part II is devoted to the development of the basic mathematical concepts of enzyme behaviour and kinetics as they affect industrial operations; Part III is based on practical data covering sources, methods of extraction, isolation and characterization of enzymes.

In the beginning of Section 'B', a chapter has been included to emphasize the importance of enzymes in food, which may serve as a bridge between general enzymology and food enzymology. Section 'B' of the book presents a comprehensive, yet concise, coverage of the latest developments in understanding the structures and properties of the major groups of enzymes, including their potential applications in the food processing industries, biotechnology and genetic engineering.

I would be grateful if I could be informed of the errors in presenting facts and interpretations, which may have been unintentionally introduced into the text.

Shahina Naz

Acknowledgements

I am deeply indebted to my teachers, in particular to late Dr Shabbir Akhtar Zubairi, Professor, Department of Chemistry, University of Karachi, for awakening my academic interests. I am also grateful to late Dr S.M. Ifzal, Ex-chairman of the Department of Food Science & Technology, University of Karachi, for giving me the opportunity to teach enzymology.

I am appreciative of the support given by Dr Syed Asad Sayeed, Assistant Professor, and Mr Rahmanullah Siddiqi, Lecturer, Department of Food Science & Technology, for their constant encouragement and invaluable assistance in the publication of this text.

My appreciation also goes to the authors whose publications are referenced in this book.

I am thankful to my family members for relieving me of my domestic duties to complete, what I hope, will be an addition to the food science library.

ENZYMES IN GENERAL

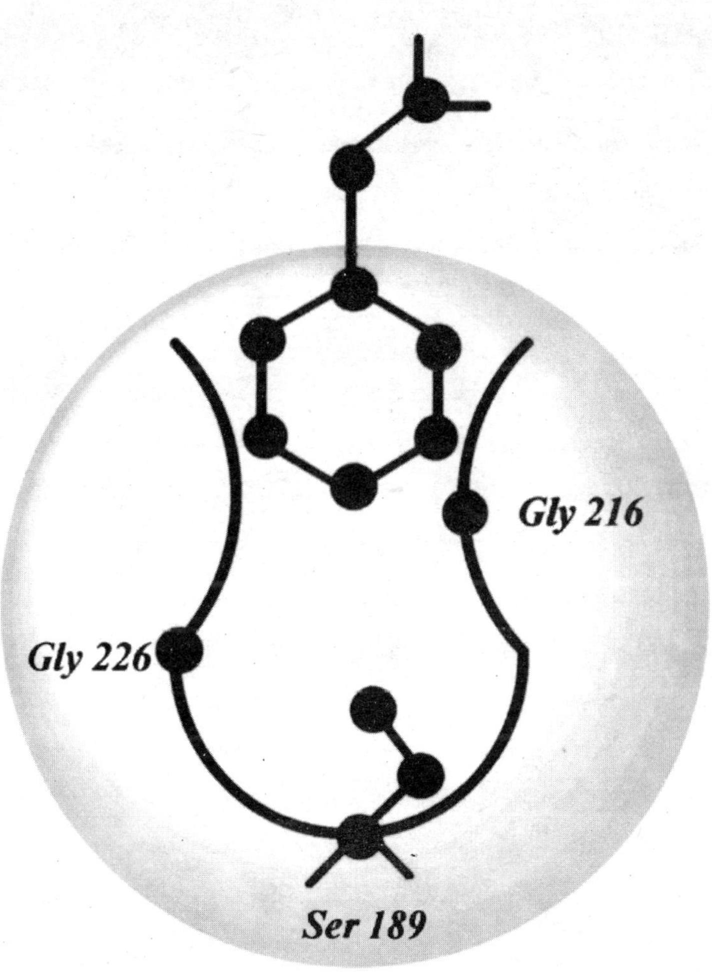

SECTION A

PART I

Enzyme as Protein

1 An Introduction to Enzymes

1.1 DEFINITION OF ENZYME AND OTHER RELATED TERMS

Enzymes are biological catalysts. They increase the rate of chemical reactions taking place within living cells without themselves suffering any overall change. The reactants of enzyme-catalysed reactions are termed substrates and each enzyme is quite specific in character, acting on a particular substrate or substrates to produce a particular product or products. The fact that enzymes act as catalysts in key biochemical reactions, makes life possible because the biochemical process known as metabolism, is characterized by chemical and physical changes continuously going on in what we call life. A majority of the biochemical reactions involved in these changes are too slow for life to keep going if enzymes are not present. The oxidation of a fatty acid to CO_2 and H_2O requires, in the laboratory, rather extreme conditions of pH and temperature, as well as, corrosive oxidizing acidic mixtures. The same chemistry, however, can occur in living cells with enzymatic help in a gentle, yet rapid manner within a narrow pH and temperature range.

All enzymes are proteins of high molecular weight (10,000 to 2,000,000), made up primarily of chains of amino acids linked together by peptide bonds. However, without the presence of a non-protein component called a cofactor, many enzyme proteins lack catalytic activity. When this is the case, the inactive component of an enzyme is termed the apoenzyme, and the active enzyme, including cofactor, the holoenzyme. The cofactor may be an organic molecule, when it is known as a coenzyme, or it may be a metal ion. Some enzymes bind cofactors more tightly than others. When a cofactor is bound so tightly that it is difficult to remove without damaging the enzyme, it is sometimes called a prosthetic group.

Substances that are able to increase the activity of the enzyme in a non-specific manner are called activators. They are part of the activating system and are required before the enzyme can activate its substrate. An activator is not a coenzyme because a coenzyme is a part of the reaction system and plays no role in the activation of the substrate. Many inorganic radicals such as chloride, potassium, calcium, magnesium and phosphates are activators.

Any compound which reduces the observed activity of an enzyme is an inhibitor. Inhibition of enzyme activity by a compound may be a result of its reaction with apoenzyme, necessary cofactors, activators, intermediates in the pathway, or the essential groups of the enzyme. The exceptional features of enzymes that make them different from non-biological catalysts include structural features, specificity towards a particular reaction or type of reaction, and mechanism of catalysis.

1.2 BRIEF HISTORY OF ENZYMES

Until the nineteenth century, it was considered that processes such as the souring of milk and fermentation of sugar to alcohol could only take place through the action of a living organism. In 1833, the active agent breaking down the sugar was partially isolated and given the name diastase (now known as amylases). Soon after, a substance which digested dietary protein was extracted from gastric juice and called pepsin. These and other active preparations were given the general name ferments. Liebig recognized that these ferments could be non-living materials obtained from living materials. It was the discovery of Liebig that replaced the term ferment by the name enzyme. This name was first proposed by Kuhne in 1878, and came from the Greek *enzume*, meaning 'in yeast'. In 1926, Summer crystallized an enzyme urease from jack-bean extracts, and in the next few years, many other enzymes were purified and crystallized. Once pure enzymes were available, their structure, function and properties could be determined.

1.3 NOMENCLATURE AND CLASSIFICATION OF ENZYMES

Earlier attempts at naming enzymes were based on the name of the discoverer, source or species from which it was isolated, method of isolation, etc. However, the first attempt at a systematic nomenclature of enzymes was the suggestion that it should be based on the name of the substrate upon which the enzyme acts and -ase be added as ending in the systematic name of enzymes. Thus, maltase is the enzyme which acts upon maltose, urease is the enzyme which acts upon urea, and amylase is the enzyme which acts upon amylose.

However, this theme of nomenclature could not keep pace with further discoveries of enzymes and substrates and faced three major difficulties. First, as more and more substrates became available for an enzyme it became more difficult to know which substrate to use in the naming. For example, an enzyme of plant, which oxidizes phenolic compounds may be given the name cresolase, catecholase, phenolase, polyphenol oxidase, etc., depending on the nature of the substrate used. Second, since the nomenclature did not indicate the nature of the chemical reaction catalysed by the enzyme, one could not tell that lactase, maltase and amylase are all enzymes, which hydrolyse glycosidic bonds in the appropriate substrate.

The last problem with the nomenclature was that it could not justify the enzymes specificity. For example, for all four enzymes—pepsin, trypsin, chymotrypsin and carboxypeptidase-A— which act on proteins, a single generic name of proteinases is sufficient to know about the type of reaction they catalyse but this name does not permit one to distinguish among these very different enzymes.

Thus, because of the lack of consistency in the nomenclature, it became apparent as the list of known enzymes rapidly grew that there was a need for a systematic way of naming and classifying them. A commission was appointed by the International Union of Biochemistry, which developed the basis of the present accepted system of nomenclature.

1.3.1 The Enzyme Commission's System of Classification

The enzyme commission divided enzymes into six main classes on the basis of the total reaction catalysed. Each enzyme was assigned a code number, consisting of four elements, separated by dots. The first digit shows to which of the main classes the enzyme belongs as:

First digit	Enzyme class
1	Oxidoreductases
2	Transferases
3	Hydrolases
4	Lyases
5	Isomerases
6	Ligases

The second and third digit in the code further describe the kind of reaction being catalysed. The meanings of these digits are defined separately for each of the main classes (Table 1.1).

Enzymes catalysing very similar but non-identical reactions, e.g. the hydrolysis of different carboxylic acid esters, will have the same first three digits in their code. The fourth digit distinguishes between them by defining the actual substrate, e.g. the actual carboxylic acid ester being hydrolysed.

However, it should be noted that isoenzymes, that is, different enzymes catalysing identical reactions, will have the same four digit classification. The classification, therefore, provides only the basis for a unique identification of an enzyme, the particular isoenzyme and its source still have to be specified.

It should also be noted that all reactions catalysed by enzymes are reversible to some extent and the classification which would be given to the enzyme for the catalysis of the forward reaction, would not be the same as that for the reverse reaction. For example, for redox reaction involving the interconversion of NADH to NAD^+, the

TABLE 1.1
Systematic basis of enzyme nomenclature

First number (Class)	Nature of reaction	Second number (First subclass)	Third number (Second class)
Oxidoreductases	Electron transfer	Group oxidized	Group reduced
Transferases	Group transfer	Group transferred (further delineated)	Group transferred
Hydrolases	Hydrolysis	Bond hydrolysed: Ester, peptide, etc.	Substrate class: Carboxylic ester, thiol ester, etc.
Lyases	Bond splitting	Bond cleaved: C-O, C-S, C-N, etc.	Group eliminated: Carboxyl, aldehyde, etc.
Isomerases	Isomerization	Type of reaction: Racemization or epimerization	Type of molecule undergoing isomerization
Ligases	Bond formation	Type of bond synthesized: C-C, C-O, C-S, C-N, etc.	Substrate, cosubstrate class

classification is usually based on the direction where NAD^+ is the electron acceptor, rather than that where NADH is the electron donor.

SUMMARY

Enzymes are proteins which catalyse, in a highly specific way, chemical reactions taking place within the living cell. Often a further, non-protein, component called a cofactor is required before an enzyme has catalytic activity.

Because of the lack of consistency and lack of clarity in the names of enzymes, an Enzyme Commission, appointed by the International Union of Biochemistry, has given all known enzymes a systematic name and a four-digit classification.

RECOMMENDATIONS

1. Dixon, M., Webb, E. C., Thorne, C.J.R. and Tipton, K.F. (1979), *Enzymes*, 3rd edn. (Chapters 1 & 5), Longman.
2. Enzyme Nomenclature Recommendations (1984) of the Nomenclature Committee of the International Union of Biochemistry. Published by Academic Press. Corrections and additions listed in *European Journal of Biochemistry* (1986), 157 (pp. 1-26) and (1989), 179 (pp. 489-553).

2 Structural Features of Enzymes

2.1 CHEMICAL NATURE OF ENZYMES

In contrast to non-biological catalysts, which may differ widely in chemical nature and structure (ranging from precious and inert metals like platinum, gold to inorganic salts like $FeCl_3$), all enzymes are proteins. Thus, knowledge of protein structure is clearly a prerequisite to any understanding of the enzyme structure.

Proteins are macromolecules with molecular weights of at least several thousand. They are found in abundance in living organisms, making up more than half the dry weight of cells. Two distinct types are known: fibrous and globular proteins.

Fibrous proteins are insoluble in water and are physically tough, which enables them to play a structural role. Examples include, α-keratin (a component of hair, nails and feathers) and collagen (the main fibrous element of skin, bone and tendon). In contrast, globular proteins are generally soluble in water and may be crystallized from solution. They have a functional role in living organisms. All enzymes are globular proteins.

TABLE 2.1

conjugated protein	extra component present
nucleoprotein	a nucleic acid
lipoprotein	a lipid
glycoprotein	an oligosaccharide
haemoprotein	an iron protoporphyrin
flavoprotein	a flavin nucleotide
metalloprotein	a metal

All proteins consist of amino acid units, joined in series. The sequence of amino acids in a protein is specific, being determined by the structure of the genetic material of the cell, and this gives each protein unique properties. Some proteins are composed entirely of these amino acid blocks and are termed simple proteins. Others, called conjugated proteins, contain extra material, which is firmly bound to one or more of the amino acid units as shown in Table 2.1.

As it has already been stated, an enzyme may be either simple or conjugated protein.

2.1.1 Amino Acids, the Building Blocks of Proteins

2.1.1.1 Structure and classification of amino acids

Amino acids, by definition, are organic compounds which contains, within the same molecule, an amino group ($-NH_2$ or $>NH$) and a carboxyl group ($-COOH$). Thus they have properties of both bases and acids.

The amino group in all of the twenty amino acids commonly found in proteins, is a primary one ($-NH_2$), the exception being proline, which contains a secondary amino group ($>NH$).

All the amino acids commonly found in proteins are α-amino acids, since the amino group is on the α-carbon atom. The general formula is:

$$H_2N - \underset{\underset{H}{|}}{\overset{\overset{R}{|}}{C}} - C \overset{\diagup O}{\diagdown OH}$$

or, more conveniently, $NH_2.CHR.COOH$

The symbol R represents the rest of the molecule, often called the side chain. The amino and carboxyl groups attached to the α-carbon atom are termed as the α-amino and α-carboxyl groups, to distinguish them from similar groups which may be present as part of the side chain.

TABLE 2.2

The side chains (–R) of the twenty amino acids commonly found in food proteins

Non-polar side chains			
Amino acid	**(–R)**	**Amino Acid**	**(–R)**
Alanine (Ala)	– CH_3	Valine (Val)	– $CH.CH_3$ | CH_3
Leucine (Leu)	– $CH_2CH.CH_3$ | CH_3	Isoleucine (Ile)	– $CH.CH_2CH_3$ | CH_3
Phenylalanine (Phe)	– CH_2 –⬡	Tryptophan (Trp)	– CH_2 (indole ring)
Methionine (Met)	– CH_2 – CH_2 – S – CH_3	Proline (Pro)	(pyrrolidine ring) $CH.CO_2^-$, $\overset{+}{N}H$

Polar side chains			
Negative Charge at pH 7			
Amino Acid	**(–R)**	**Amino Acid**	**(–R)**
Aspartic acid (Asp)	– $CH_2C\!\!\begin{smallmatrix}O\\\\O^-\end{smallmatrix}$	Glutamic acid (Glu)	– $CH_2CH_2C\!\!\begin{smallmatrix}O\\\\O^-\end{smallmatrix}$
Positive Charge at pH 7			
Amino Acid	**(–R)**	**Amino Acid**	**(–R)**
Lysine (Lys)	– $(CH_2)_4\overset{+}{N}H_3$	Arginine (Arg)	– $(CH_2)_3NHC.NH_2$, $\|$, $\overset{+}{N}H_2$
Uncharged at pH 7			
Amino Acid	**(–R)**	**Amino Acid**	**(–R)**
Glycine (Gly)	– H	Serine (Ser)	– CH_2OH
Threonine (Thr)	– $CH.CH_3$ | OH	Cysteine (Cys)	– CH_2SH
Tyrosine (Tyr)	– CH_2 –⬡– OH	Asparagine (Aspn)	– $CH_2C\!\!\begin{smallmatrix}O\\\\NH_2\end{smallmatrix}$
Glutamine (Glun)	– $CH_2CH_2C\!\!\begin{smallmatrix}O\\\\NH_2\end{smallmatrix}$	Histidine (His)	– CH_2 (imidazole ring)

Amino acids with non-polar or hydrophobic side chains include alanine, isoleucine, leucine, methionine, phenylalanine, proline, tryptophan, and valine. They are less soluble in water than polar amino acids. Their hydrophobicity increases with the length of the aliphatic side chain. Amino acids with polar, uncharged (hydrophilic) side chains possess neutral, polar, functional groups which are able to establish hydrogen bonds with appropriate molecules, such as water. The polarity of serine, threonine and tyrosine is related to their hydroxyl groups (-OH), the polarity of asparagine and glutamine to their amide group (-CO-NH_2), and the polarity of cysteine to its thiol group (-SH). Glycine is sometimes included in this class. Cysteine and tyrosine possess the most polar functional groups in this class, since both thiol and phenol groups may undergo partial ionization at pH values close to neutrality. In proteins, cysteine is often present in an oxidized state, that is, as cystine. This occurs when the thiol groups of two cysteine molecules oxidise and form a disulphide cross-link. Amino acids with positively charged side chains (at a pH close to 7) consist of lysine, arginine and histidine. Amino acids with negatively charged side chains (at a pH close to 7) consist of aspartic and glutamic acids.

Thus, amino acids with a considerable variety of side chain characteristics are found in proteins. This explains the range of properties shown by these macromolecules.

2.1.1.2 Stereochemistry of natural amino acids

If we consider the bonds involving the α-carbon of an amino acid, we see that two different spatial arrangements, or stereoisomeric forms, are possible. α-carbon is covalently linked to four different atoms or groups, so it is asymmetric. As a consequence of this, two mirror image forms of the molecule can exist. Such forms are termed optical isomers, since one (D-amino acid) will rotate the plane of polarized light passing through it to the right, and the other (L-amino acid) to the left.

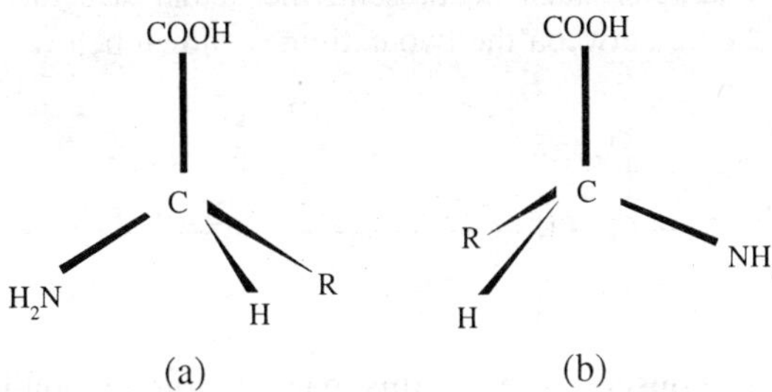

Fig. 2.1 *Three dimensional arrangement about the α-carbon atom for (a) an L-amino acid and (b) D-amino acid.*

Except for glycine, all the natural amino acids resulting from mild protein hydrolysis (acid or enzymic) exist as optical isomers. Glycine does not have the asymmetric carbon atom since in this case there are two hydrogen atoms attached to the α-carbon (R=H). Four amino acids—isoleucine, threonine, hydroxylysine and hydroxyproline—have a second asymmetry center and therefore, each one has four stereoisomers.

2.2 THE LEVELS OF PROTEIN STRUCTURE

2.2.1 Primary Level of Protein Structure

Four separate levels of protein structure can be determined—the primary, secondary, tertiary and quaternary structures. The primary structure is the sequence of amino acids making up the protein, a peptide bond connects the α-carboxyl group of each amino acid to the α-amino group of the next in the chain.

$$H_2N.CHR^1.CO_2H + H_2N.CHR^2.CO_2H \rightarrow$$

$$H_2N.CHR^1.\overset{\overset{\textstyle O}{|}}{C}-\overset{\overset{\textstyle H}{|}}{N}.CHR^2.CO_2H + H_2O$$

Since a molecule of water is lost when two free acid molecules undergo this reaction, only their residues are linked. A molecule consisting of two amino acid residues joined by a peptide bond is called a dipeptide. Several residues linked in this way form an oligopeptide, while a chain of many amino acid residues is termed a polypeptide. The covalent backbone of such a structure consists of α-carbon atoms linked by peptide bonds, the R groups sticking out from the chain. Each peptide chain has one free amino end (the N-terminus) and one free carboxyl end (the C-terminus); all the other α-amino and α-carboxyl groups present are involved in peptide bonds. For example:

N-terminus   C-terminus

$$N_2H.CHR^1.CONH.CHR^2.CONH.CHR^3.CONH.CHR^4.CO_2H$$

N-terminal   C-terminal
amino acid residue   amino acid residue

Proteins may contain one or more polypeptide chains, each one having a specific primary structure. Although, a two-dimensional representation of a polypeptide chain can give the impression that the backbone is linear, it should be

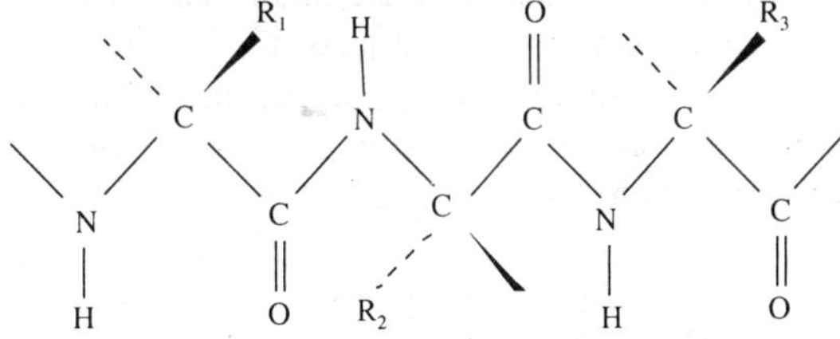

Fig. 2.2 *Structure of a fragment of an α-L-polypeptide chain (in the trans configuration).*

understood that this is not so. The even distribution in three-dimensional space of the single covalent bonds about the carbon and nitrogen atoms in the backbone means, that no two bonds emerging from the same atom will be diametrically opposite to each other. Molecules may rotate freely about single covalent bonds, so an unlimited number of arrangements of a polypeptide chain in space is possible. However, some of these will be more stable than others, so are more likely to exist.

2.2.2 Protein Structure beyond Primary Level

Based on primary structure alone, a protein would be an exceedingly long, thick molecule that would flop around in solution in every direction as a random coil and become enmeshed with itself and other polymeric molecules in solution. For example, ribonuclease with 124 amino acids would be 448 A^0 long and about 3.7 A^0 thick. Since this is not so, it can easily be stated that there must be additional structures imposed upon the primary structure to cause the protein to be so compact. These additional structures are nothing more than the various types of interactions and linkages that develop between the amino acids of the same polypeptide chain or between amino acids of the two different and adjacent polypeptide chains. The types of interactions and linkages involved in the formation of the additional structures include Vander Waals interactions, electrostatic interactions, hydrogen bonding, and hydrophobic interactions. The higher and multiple level of the protein structure and their degree of compactness is a direct consequence of the nature and number of various bonds that connects the amino acid residues within the same polypeptide chain or amino acid residues belonging to two different polypeptide chains.

2.2.2.1 Secondary level of protein structure

Since in a polypeptide chain the substituents of the α-carbon of each amino acid can rotate around the axes constituted by single covalent bonds, there are numerous possibilities for the spatial structure of a polypeptide chain.

The secondary structure of a protein is the spatial structure, the polypeptide chain assumes exclusively along the axis. Much of our understanding of the secondary structure of proteins is the result of X-ray analysis which shows that:

• C-N bond length on a peptide bond is much shorter than can be expected for a single covalent bond. Therefore, some degree of double bond character must be present, the actual structure being between the two extremes shown below:

A consequence of this partial double bond character is that rotation about the bond is restricted and all the atoms involved lie in the same plane. Two isomeric arrangements are possible: the trans form, with the oxygen and hydrogen atoms diametrically opposed, and the cis form, with these atoms adjacent. In fact, only the trans isomer is found.

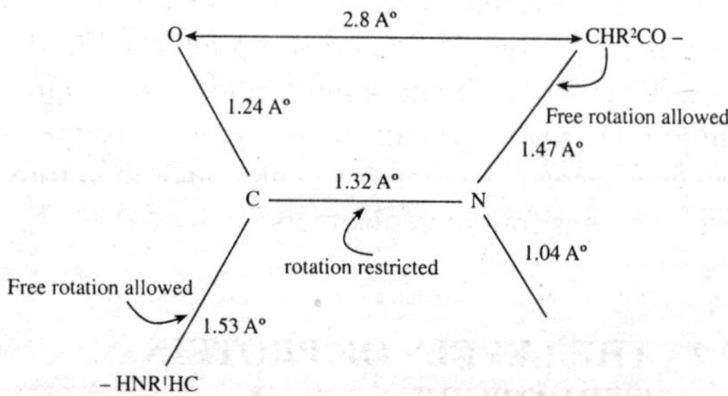

Fig. 2.3 *The dimensions of the peptide bond.*

• The most significant factor accounting for the stability of the trans form is the spacing between the α-carbon atom and the oxygen atom (2.8 A^0) which is only marginally less than the Vander Waals contact distance between these atoms (3.4 A^0), so repulsion is slight. In the unstable cis form, the two α-carbon atoms would be adjacent to each other, separated by a distance (2.8 A^0), much less than the Vander Waals contact distance between two carbon atoms (4.0 A^0), hence, repulsive forces would be greater.

- There is a high degree of hydrogen bonding between the oxygen atoms of one peptide bond and nitrogen atoms of the other. The distance between such atoms is often about 2.9 A^0, much less than the Vander Waals contact distance between non-bonded oxygen and nitrogen atoms, thus indicating the presence of the hydrogen bond. Furthermore, the N-H—O linkage is usually approximately linear.
- For many proteins the X-ray diffraction pattern indicates a regular repetition of certain structural units.

On the basis of these findings and the fact that all amino acids are of L-configuration, various theoretical types of secondary structure have been proposed.

2.2.2.1a Flat-sheet structure

Initially, to understand all possible types of secondary structures, it is convenient to start with a structure in which peptide chains are fully extended to form flat zig-zags:

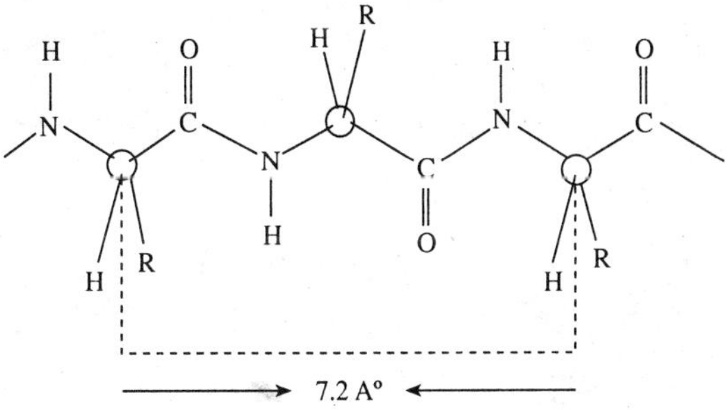

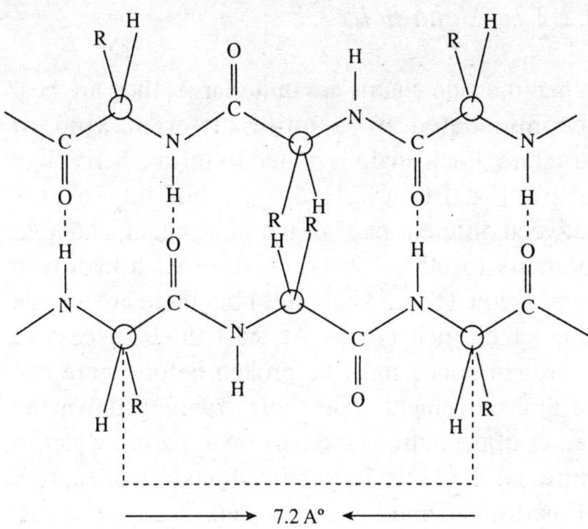

Fig. 2.5 *Hypothetical flat sheet structure for a protein.*

2 2.2.1b Beta-pleated sheets

Room can be made for small or medium-sized side chains by a slight contraction of the peptide chains. The chains still lie side by side, held to each other by hydrogen bonds. The contraction results in a pleated sheet, with a somewhat shorter distance between alternate amino acid residues. Such a structure, called the beta arrangement, has a repeat distance of 7.0 A^0. Enzymes contain variable amounts of α-pleated sheet structure.

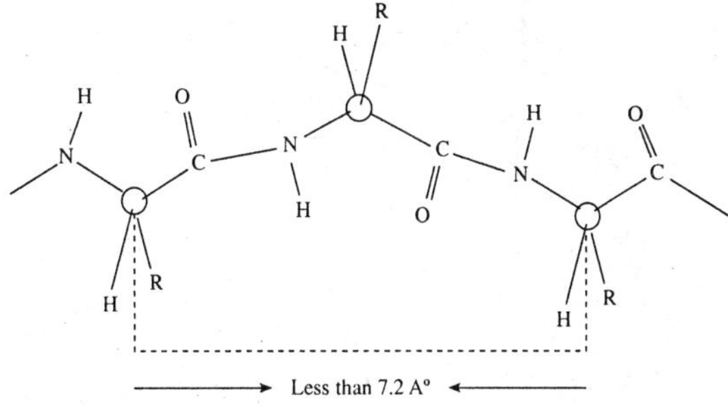

Fig. 2.6 *Contracted peptide chain.*

Fig. 2.4 *Extended peptide chain.*

These chains lie side by side to form a flat sheet. Each chain is held by hydrogen bonds to the two neighbouring chains. This structure has a repeat distance of 7.2 A^0, the distance between alternate amino acid residues. However, crowding between side chains makes this idealized flat structure impossible.

2.2.2.1c Alpha-helix

When the side chains are quite large, they are best accommodated by a quite different kind of structure. Each chain is coiled to form a helix (like a spiral staircase). Hydrogen bonding occurs between different parts of the same chain, and hold the helix together. The strength of each hydrogen bond is low (1 to 5 kcal/mole) but there are a large number of such bonds. At least three successive hydrogen bonds must be broken before there can be any movement. The 'hole' running down the center of the helix is too narrow to permit water to entre, so the helix is stabilized against disruption of hydrogen bonding by agents such as water.

the distance between amino acid residues measured along the axis of the helix. To fit into this helix, all the amino acid residues must be of the same configuration, as, of course, they are; furthermore, their L-configuration requires the helix to be right handed, as shown. Such a structure, called the alpha-helix, is of fundamental importance in the chemistry of proteins.

2.2.2.2 Tertiary level of protein structure

The picture we have developed so far of a protein is that it consists of a long polypeptide backbone distributed in segments of alpha-helices, and random coils. Again, this is not sufficient to account for the compact, near-spherical nature of most enzymes. X-ray diffraction patterns of proteins show that the polypeptide chain is folded back on itself in a unique and reproducible fashion.

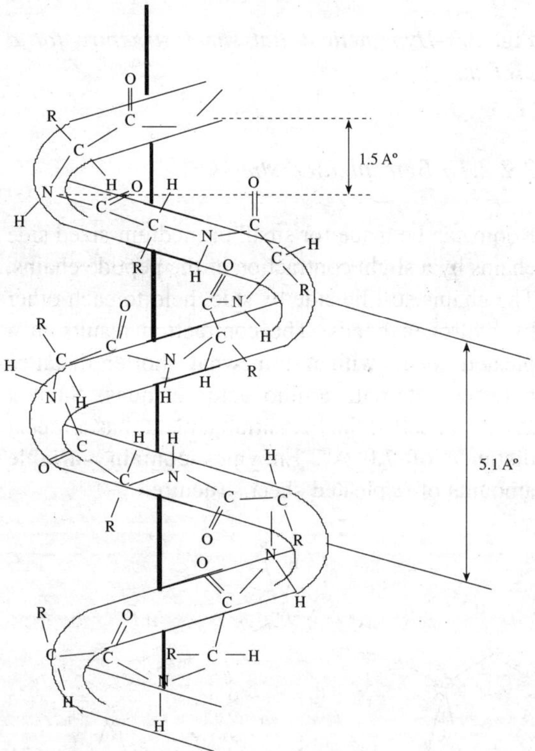

Fig. 2.7 *Alpha-helix structure proposed by Pauling.*

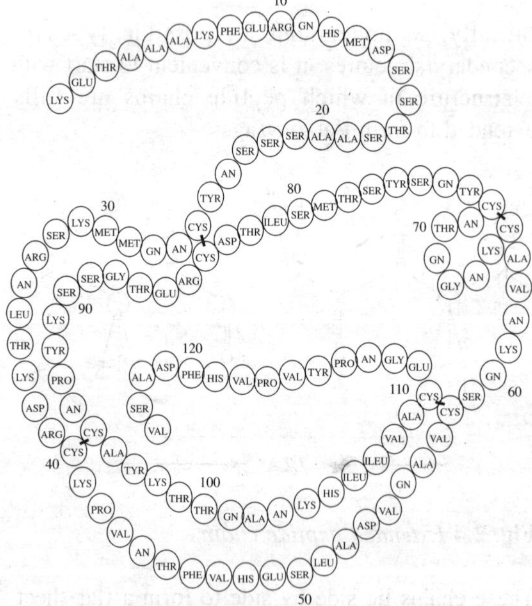

Fig. 2.8 *Primary structure of ribonuclease (AN, asparagine; GN, glutamine). Reprinted with permission, © Academic Press.*

There are 3.6 amino acid residues per turn. This 3.6-helix provides room for the side chains and allows all possible hydrogen bonds to form. It accounts for a repeat distance of 1.5 A^0, which is

In myoglobin, 118 of the 153 amino acid residues are involved in eight helical segments of lengths ranging from 7 to 24 amino acid residues each. The α-helical segments are separated by two sharp turns and five non-helical regions of 1 to 8 amino acid residues. There is another non-helical region involving the 5 amino acid residues at the carboxyl terminal end of the chain. In the tertiary structure the protein is folded in a complex and irregular manner to form a compact structure. This folding is such that the haem group and almost all of the polar amino acid residues are on the surface and almost all of the non-polar residues are in the interior of the molecule. Thus, the hydrophilic residues are exposed to the solvent, water, while the hydrophobic residues are removed from the water as much as possible.

The types of bonds involved in maintaining the tertiary structure of proteins include electrostatic bonds, hydrogen bonds, hydrophobic bonds, dipolar bonds and disulphide bonds. Not every protein has all of these types of bonds involved in maintaining its tertiary structure.

The reproducible nature of the folding or tertiary structure has been shown by unfolding proteins and permitting them to refold. When the four disulphide bonds in ribonuclease are reduced in urea solution, the polypeptide chain assumes a random coil conformation and loses all enzymatic activity. When the urea and reducing agents are removed and the disulphide bonds are allowed to reform slowly by oxidation in air, more than 90 per cent of the enzymatic activity is regained and the properties of the molecule are in every way identical to those of the original molecule. This is quite astonishing as the disulphide bonds could have been reformed at random to give 105 different forms of the ribonuclease. Truly, the primary amino acid sequence must determine the higher structures. The higher structures of a protein must be considered to be as unique as is the primary structure.

2.2.2.3 Quaternary level of protein structure

Many proteins are composed of a single polypeptide chain. Monomeric proteins are those which consist of only a single polypeptide chain, so they cannot be dissociated into smaller units. Very few monomeric enzymes are known, and all of these catalyse hydrolytic reactions. In general they contain between 100 and 300 amino acid residues and have molecular weights in the 13,000–35,000 range. A number of proteases (or proteolytic enzymes) such as trypsin, chymotrypsin, pepsin, papain, ficin, and some amylases are monomeric.

Oligomeric proteins consist of two or more polypeptide chains, which are usually linked to each other by non-covalent interactions and never by peptide bonds. The component polypeptide chains are termed sub-units and may be identical to or different from each other, if they are identical, they are sometimes called protomers. Dimeric proteins consist of two, trimeric proteins of three and tetrameric proteins of four sub-units. The quaternary structure of a protein refers to a complete three-dimensional structure, including the interactions between the component polypeptide chains which may or may not be identical.

The majority of known enzymes are oligomeric. The molecular weight is usually in excess of 35,000. All of the enzymes involved in glycolysis possess either two or four sub-units. Lactate dehydrogenase is composed of four polypeptide chains of two different types. The monomeric units are inactive until assembled into the tetramer. Aspartate transcarbamylase is composed of two different types of polypeptide chains. In the intact molecule there are six polypeptide chains of each type. One type of polypeptide chain binds substrate and catalyses the reaction of carbamyl phosphate and L-aspartate to form carbamylaspartate, while the other type of polypeptide is inactive as an enzyme but binds the allosteric inhibitor cystidine triphosphate. These two types of chains are referred to as catalytic polypeptides and regulatory polypeptides respectively. The two types of sub-units can be separated but after separation, though the catalytic units fully retain their activity, there is no allosteric behaviour of the enzyme in the absence of the regulatory sub-units. It is, therefore, reasonable to assume that the sub-units of oligomeric proteins gain properties in association that they do not have in isolation.

SUMMARY

Proteins consist of L-amino acid residues linked by peptide bonds. The sequence of amino acids in each polypeptide chain constitutes the primary structure of the protein. Regular, repeated three-dimensional features constitute the secondary structure. This is largely uninterrupted in fibrous, structural proteins but disrupted at many points in globular and functional proteins, including enzymes. The overall three-dimensional structure of each polypeptide chain is termed the tertiary structure. Proteins may consist above one or more polypeptide chains, the complete structure being called the quaternary structure.

RECOMMENDATIONS

1. Colowick, S.P. and Kaplan, N.O. (eds.), *Methods in Enzymology*, 47 (1977), 48, 49 (1978), 61 (1979), 91 (1983), 130, 131 (1986); *Enzyme structure*; 114, 115 (1985), Academic Press.
2. Eisenberg, D. and Hill, C.P. (1989), 'Protein Crystallography', *Trends in Biochemical Sciences*, 14 (pp. 260-64).
3. Hunkapiller, M.W., Strickler, J.E. and Wilson, K.J. (1984), 'Contemporary Methodology for Protein Structure Determination', *Science*, 226 (pp. 304-11).
4. Price, N.C. and Stevens, L. (1989), *Fundamentals of Enzymology*, 2nd edn. (Chapter 3), Oxford University Press.
5. Shivey, J.E., Paxton, I.J. and Lee, T.D. (1985), 'Highlights of Protein Structural Analysis', *Trends in Biochemical Sciences*, 14 (pp. 246-52).
6. Voet, D. and Voet, J.G. (1990), *Biochemistry* (Chapters 4-7), John Wiley.
7. Wilson, K. and Goulding, K.H. (1986), *Principles and Techniques of Practical Biochemistry*, 3rd edn. (Chapter 8), Edward Arnold.
8. Wright, P.E. (1989), 'What can 2-D NMR tell us about Proteins ?' in *Trends in Biochemical Sciences*, 14 (pp. 255-60).
9. Zubay, G. (1988), *Biochemistry*, 2nd edn. (Chapters 1-3), Macmillan.

3 Biosynthesis and Properties of Proteins

3.1 BIOSYNTHESIS OF PROTEINS

The fact that amino acid sequence of each globular protein in a given species is absolutely specific implies that the biosynthesis of proteins is under genetic control and hence, connected with nucleic acid structure and function.

3.1.1 Nucleic Acids

Although chemically quite different, nucleic acids resemble proteins in a fundamental way. There is a long chain, 'a backbone', that is the same (except for length) in all nucleic acid molecules and attached to this backbone are various groups, which by their nature and sequence characterize each individual nucleic acid.

Where the backbone of the protein molecule is a polyamide chain (a polypeptide chain), the backbone of the nucleic acid molecule is a polyester chain (called a polynucleotide chain). The ester is derived from phosphoric acid (the acid portion) and a sugar (the alcohol portion).

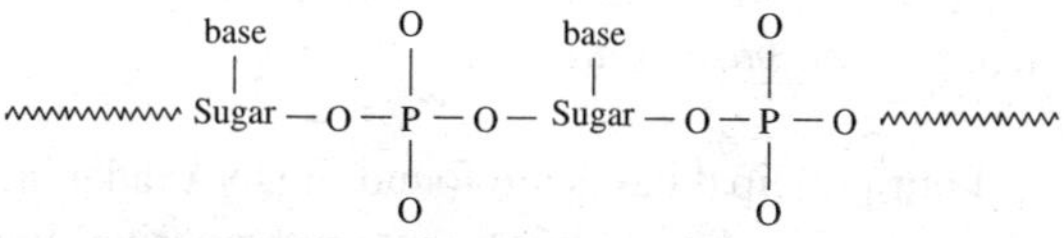

Fig. 3.1 *Polynucleotide chain.*

The sugar is D-ribose in the group of nucleic acids known as ribonucleic acids (RNA) and D-2-deoxyribose in the group known as deoxyribonucleic acids (DNA). The sugar units are in the furanose form and are joined to phosphate through the C-3 and C-5 hydroxyl groups (Figure 3.2).

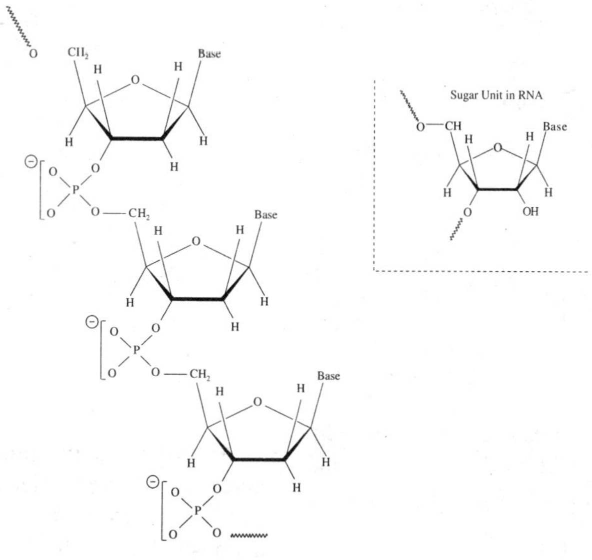

Fig. 3.2 *Deoxyribonucleic acid (DNA) and Ribonucleic acid (RNA).*

Attached to C-1 of each sugar is one of a number of heterocyclic bases. A base-sugar unit is called a nucleoside, a base-sugar-phosphoric acid unit is called a nucleotide. An example of a nucleotide is shown in Figure 3.3.

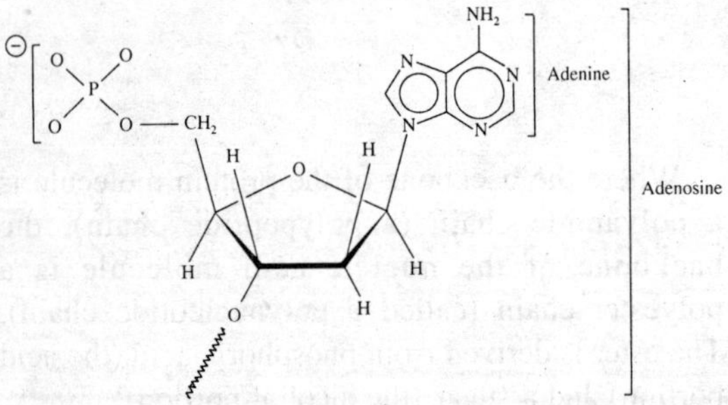

Fig. 3.3 *A nucleotide: an adenylic acid unit of RNA. Here the nucleoside is adenosine and the heterocyclic base is adenine.*

Four principal bases are found in DNA, adenine (A) and guanine (G) which contain the purine ring system, and cytosine (C) and thymine (T) which contain the pyrimidine ring system. RNA contains adenine, guanine, cytosine and uracil (U).

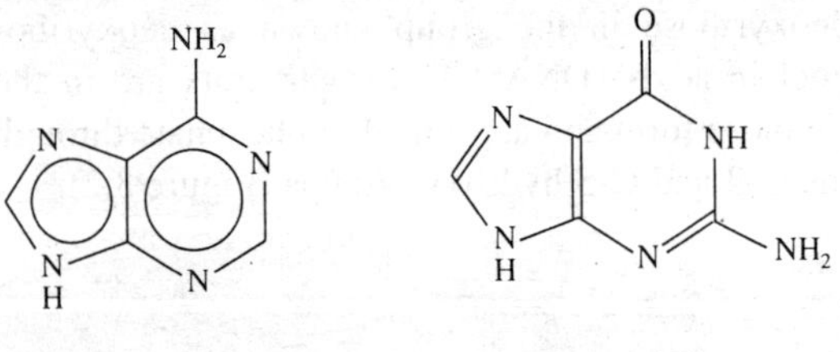

Adenine (A)   **Guanine (G)**

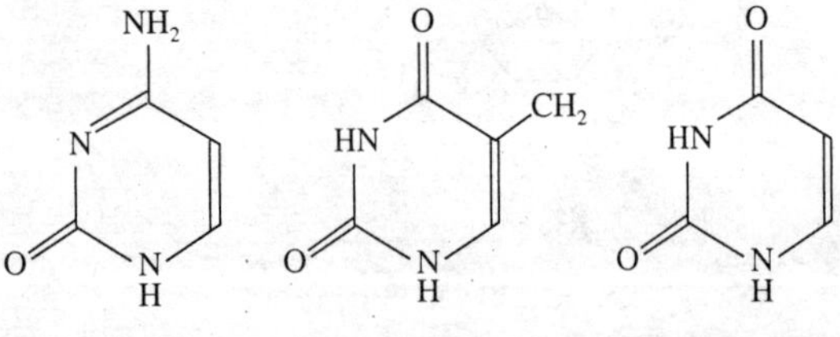

Thymine (T)   **Uracil (U)**   **Cytosine (C)**

The proportions of these bases and the sequence in which they follow each other along the polynucleotide chain differ from one kind of nucleic acid to another.

3.1.2 Double Helix of DNA

Like proteins, after evaluation of primary structure, the problem was to devise a secondary structure for nucleic acids which could account for the chemical and X-ray evidence and at the same time be consistent with all the structural features of the units involved, molecular size and shape, bond angles and bond lengths and configurations and conformations. Of the chemical evidences the most valuable clue was this: although the proportions of bases vary from one DNA to another, A is always bonded to T and G is always bonded to C.

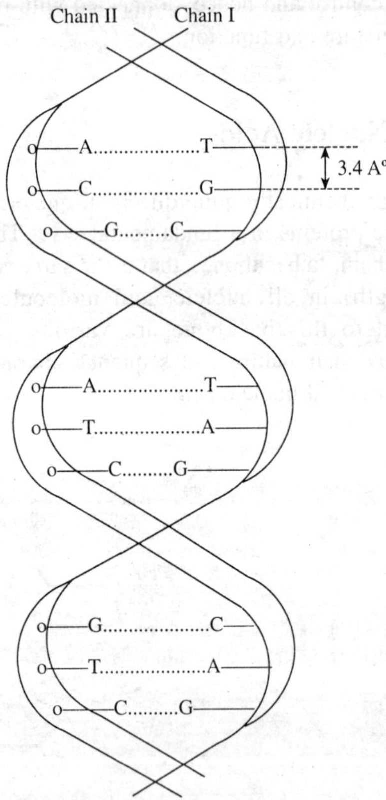

Fig. 3.4 *Double helix structure proposed for DNA.*

According to the proposed structure DNA is made up of two nucleotide chains wound about each other to form a double helix 20 A° in diameter. Each helix is right-handed and has ten nucleotide units for each complete turn, which occurs every 34 A⁰ along the axis. The two chains head in opposite directions, that is, the deoxyribose units are oriented in opposite ways so that the sequence is C-3, C-5 in one chain and C-5, C-3 in the other.

The chains are held together at intervals by hydrogen bonds. These are linear hydrogen bonds between adenine and thymine and between guanine and cytosine. As stated before, A is always bonded to T and G is always bonded to C, hydrogen bonding between other pairs of bases would not allow them to fit into the double helical structure. The two strands are thus not identical but complementary: opposite every A of one chain is a T in the other, and opposite every G is a C. Unlike DNA, secondary structure of RNA always involves single-strand helixes.

Now, there is a matter of guiding the synthesis of proteins. A particular sequence of bases along a polynucleotide chain leads to a particular sequence of amino acid residues along a polypeptide chain. A protein is like a long sentence written in a language of 20 letters, the 20 amino acid residues. But the message for its synthesis is written in a language of only four letters, A, G, T, and C. It is written in a code, with each word standing for a particular amino acid.

3.1.3 Translation of Genetic Information into Protein Structure

In protein synthesis, the double helix of DNA partially uncoils and about one of the separated strands replicates to form a chain of RNA. The base sequence along the RNA chain is different from that along the DNA template but is determined by it. Opposite to each adenine of DNA, there appears on RNA, a uracil; opposite guanine, cytosine; opposite thymine, adenine; opposite cytosine, guanine. Thus AATCAGTT on DNA becomes UUAGUCAA on RNA.

One kind of RNA, called messenger RNA, carries a message to the ribosome where protein synthesis actually takes place. At the ribosome, messenger RNA calls up a series of transport RNA molecules each of which is loaded with a particular

amino acid. The sequence in which the amino acids are built into the protein chain is dependent upon the order in which the transport RNA molecules are called up and this in turn depends upon the sequence of bases along the messenger RNA chain. Thus GUA is the code for aspartic acid; UUU for phenylalanine and GUG for valine. There are 64 three-letter code for words (codons) and only 20 amino acids, so that more than one codon can call up the same amino acids: CUU and UCU for leucine, GAA and GAG for glutamic acid.

Thus, the structure of nucleic acid molecules determines the structure of proteins and thus of enzymes.

3.2   PROPERTIES OF PROTEINS

3.2.1 Chemical Properties of Proteins

The chemical properties of proteins are largely those of the side chains of the constituent amino acids. Thus, arginine side chains, each containing a guanidine group, can react with α-naphthol in the presence of an oxidizing agent, such as sodium hypochlorite to produce a red colour: this is the Sakaguchi reaction. Similarly, tryptophan side chains, being indoles, can react with glyoxylic acid in the presence of concentrated sulphuric acid to produce a purple colour: this is the Hopkins-Cole reaction.

Tyrosine side chains, each possessing a phenolic group, can undergo a variety of reactions. If treated with mercuric sulphate and sodium nitrate and then heated, a red complex is produced by the Millon reaction. They also undergo the Folin-Ciocalteu reaction if treated with tungstate and molybdate, a blue colour being formed.

All of these procedures, particularly the last mentioned, can be used for the quantitative estimation of proteins, the intensity of the colour produced being dependent on the number of reacting groups present. However, it is usually necessary to assume that the protein being estimated has an average distribution of amino acid residues, and no reacting groups other than those in the protein must be present.

An alternative method for the quantitative estimation of protein is the biuret reaction, treatment with cupric sulphate in alkali gives a purple complex, the reacting unit in this case is the peptide bond, so free amino acids do not react. The widely used Lowry method for protein determination combines the biuret and Folin-Ciocalteu procedures. Bicinchoninic acid may also be used to detect cuprous ions formed during the biuret reaction, an intense purple complex being formed. A further method for protein analysis makes use of the fact that tyrosine and tryptophan side chains absorb light at 280nm.

Functional groups in the amino acid side chains play an important role in the catalytic activity of enzymes. Many agents can inactivate enzymes by binding to these functional groups, for example, heavy metal ions (e.g. Ag^+) bind strongly to the sulphahydryl groups of cysteine residues and thus, may act as poisons to a great many enzymes.

$$E\text{-}SH + Ag^+ \longrightarrow E\text{-}S\text{-}Ag + H^+$$

3.2.2 Acid-Base Properties of Proteins

Although the amino acids are commonly shown as containing an amino group and a carboxyl group, $H_2NCHRCOOH$, certain properties, both physical and chemical, are not consistent with this structure.

(a) In contrast to amines and carboxylic acids, the amino acids are non-volatile crystalline solids which melt with decomposition at fairly high temperatures.

(b) They are insoluble in non-polar solvents like petroleum ether, benzene, or ether, and are appreciably soluble in water.

(c) Their aqueous solutions behave like solutions of substances of high dipole moment.

(d) Acidity and basicity constants are extremely low for -COOH and $-NH_2$ groups.

All these properties are quite consistent with a dipolar ion structure for the amino acids (I).

$$^+H_3N\text{-}CHR\text{-}COO^- \text{ (I)}$$
Amino acid: dipolar ions

When the solution of an amino acid is made alkaline, the dipolar ion (I) is converted into the anion (II); Hydroxide ion, being the stronger base, removes a proton from the ammonium ion and displaces the weaker base, the amine.

$$^+H_3N\text{-}CHR\text{-}COO^- + OH^- \rightleftharpoons H_2N\text{-}CHR\text{-}COO^- + H_2O$$

Stronger acid   Stronger base   Weaker base Weaker acid
   (I)   (II)

When the solution of an amino acid is made acidic, the dipolar ion (I) is converted into the cation (III); Oxonium ion, being the stronger acid, H_3O^+, gives up a proton to the carboxylate ion and displaces the weaker carboxylic acid.

$$^+H_3N\text{-}CHR\text{-}COO^- + H_3O^+ \rightleftharpoons {}^+H_3N\text{-}CHR\text{-}COOH + H_2O$$

Stronger base   Stronger acid   Weaker acid Weaker base
   (I)   (III)

Thus acidic group of a simple amino acid like glycine, is NH_3 not ^+COOH, and the basic group is $-COO^-$ not $-NH_2$.

It must be kept in mind that ions (II) and (III), which contain a free $-NH_2$ or -COOH group, are in equilibrium with dipolar ion (I); consequently, amino acids undergo reactions characteristic of amines and carboxylic acids.

$$H_2N\text{-}CHR\text{-}COO^- \rightleftharpoons {}^+H_3N\text{-}CHR\text{-}COO^- \rightleftharpoons {}^+H_3N\text{-}CHR\text{-}COOH$$

  (II)   (I)   (III)

What happens when a solution of an amino acid is placed in an electric field, depends upon the acidity or basicity of the solution. In quite alkaline solution, anions (II) exceed cations (III), and there is a net migration of amino acid toward the anode. In quite acidic solution, cations (III) are in excess, and there is a net migration of amino acid toward the cathode. If (II) and (III) are exactly balanced, there is no net migration; under such conditions, any one molecule exists as a positive ion and as a negative ion for exactly the same amount of time, and any small movement in the direction of one electrode is subsequently cancelled by an equal movement back toward the other electrode. The hydrogen ion concentration of the solution, in which a particular amino acid does not migrate under the influence of an electric field, is called the isoelectric point (pI) of that amino acid.

A monocarboxylic acid, $^+H_3NCHRCOO^-$ is somewhat more acidic than basic (for example glycine). If crystals of such amino acid are added to water, the resulting solution contains more of the anion (II), $H_2NCHRCOO^-$, than of the cation (III), $^+H_3NCHRCOOH$. This 'excess' ionization of ammonium ion to amine (I $\rightleftharpoons$ II + H$^+$) must be repressed by the addition of an acid, to reach the isoelectric point, which therefore lies somewhat on the acidic side. For example, the isoelectric point for glycine is at pH 6.1.

An amino acid usually shows its lowest solubility in a solution at the isoelectric point, since here is the highest concentration of the dipolar ion. As the solution is made more alkaline or more acidic, the concentration of one of the more soluble ions, (II) or (III), increases.

Different proteins have different proportions of acidic and basic side chains, and hence have different isoelectric points. In a solution of particular hydrogen ion concentration, some proteins move toward a cathode and others toward an anode, depending upon the size of the charge, as well as upon molecular size and shape, different proteins move at different speeds. This difference in behaviour in an electric field is the basis of one method of separation and analysis of protein mixtures called electrophoresis.

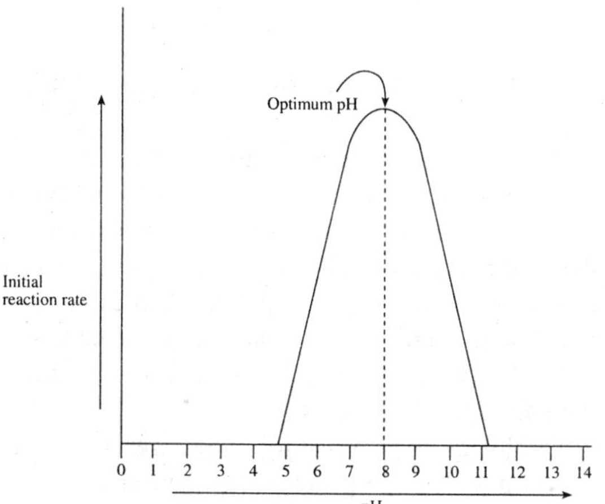

Fig. 3.6 *Effect of pH on enzyme activity.*

Globular proteins often function correctly only when certain ionizable side chains are in a specified form, making their usefulness pH-dependent. Each enzyme, therefore, has a characteristic pH optimum and is active over a relatively small pH range, in many cases a bell-shaped plot of activity against pH is obtained.

3.2.3 Solubility of Proteins

The water solubility of proteins is a function of numerous parameters. From a thermodynamic standpoint, solubilization corresponds to separating the molecules of solvent, separating the molecules of proteins, and dispersing the latter in the solvent with maximum interaction between the protein and solvent. To be soluble, a protein should therefore be able to interact as much as possible with the solvent (hydrogen-bond, dipole-dipole, and ionic interactions).

The extent of interaction of the protein molecules with solvent in turn is determined by its molecular shape. The difference in water solubility of globular and fibrous proteins is related to difference in molecular shape, which is indicated in a rough way by their names. Molecules of fibrous proteins are long and thread-like, and tend to lie side by side to form fibers. They are held together at many points by hydrogen bonds. As a result, the intermolecular forces that must be overcome by a solvent are very strong.

Molecules of globular protein are folded into compact units that often approach spherical shapes. The folding takes place in such a way, that the lipophilic parts are turned inward, toward each other, and away from water; hydrophilic parts tend to stay on the surface where they are near water. Hydrogen bonding is chiefly intramolecular. Areas of contact between molecules are small, and intermolecular forces are comparatively weak. Other factors that influence solubility of proteins are pH, ionic strength, the type of solvent, and temperature.

3.2.3.1 Influence of pH

At pH values higher or lower than the isoeletric point, the protein carries a negative or positive electric charge and water molecules may interact with these charges, thus contributing to solubilization. Moreover, protein chains carrying electrical charges of the same sign have a tendency to repel each other and to dissociate or unfold. If the solubility of a given protein is plotted as a function of the pH, one usually obtains a V- or U-shaped curve, the minimum of which corresponds closely to the pI. This behaviour is put to use for dissolving a number of proteins, especially seed proteins.

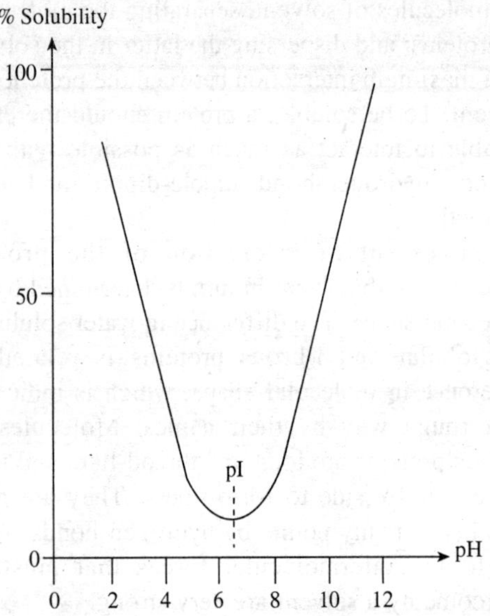

Fig. 3.7 *Effect of pH on solubility of protein.*

For pH values not far from the pI, protein molecules show minimal interactions with water and their net charges are sufficiently small to allow polypeptide chains to approach each other. Sometimes aggregates are formed, and this may lead to protein precipitation. The rate of precipitation is enhanced when the bulk densities of the aggregates differ greatly from that of the solvent.

3.2.3.2 Influence of ionic strength (μ)

The ions of neutral salts, at molarities of the order of 0.5-1M, may increase the solubility of proteins.

This effect is called 'salting in'. The ions react with the charges of proteins and decrease the electrostatic attraction between opposite charges of neighbouring molecules. Moreover, the solvation connection with these ions serves to increase the solvation of the proteins and thereby increase their solubility.

If the concentration of neutral salts is greater than 1M, the protein displays a decreased solubility, which may result in precipitation. This 'salting out' effect results from the competition between the protein and the salt ions for the water molecules necessary for their respective solvations. At high salt concentrations, there are not enough water molecules available for protein solvation, since the majority of the water molecules are strongly bound to the salts. Thus protein-protein interactions become more powerful than protein-water interactions and this may lead to aggregation followed by precipitation of the protein molecules. Some cations, e.g. Zn^{2+}, Pb^{2+}, have a more direct action on protein solubility, linking with protein anions to form insoluble complexes. Proteins may also be precipitated by treatment with various acids, e.g. trichloroacetic acid, perchloric or picric acids, which form acid-insoluble salts with protein cations. Such techniques are often used to remove proteins from solutions prior to the analysis of other substances.

3.2.3.3 Influence of non-aqueous solvents

Certain solvents, such as ethanol or acetone, lower the dielectric constant of the aqueous medium in which a protein is dissolved. As a consequence, the electrostatic forces of repulsion among protein molecules decrease and this contributes to their aggregation and precipitation. These solvents also compete for water molecules, thus further reducing the solubility of proteins.

Techniques involving varying pH, salt concentration and organic solvent concentration are widely used to separate mixtures of proteins by differential precipitation.

3.2.3.4 Influence of temperature

The solubility of globular proteins increases with the temperature upto about 40-50°C. Above this temperature, thermal agitation tends to disrupt tertiary structure, leading to denaturation and a sharp decrease in solubility. In enzymes, this effect is paralleled to some extent by changes in activity: rates of enzyme-catalysed reactions increase with increasing temperature, since collisions between molecules become more frequent, until the enzyme is denatured and catalytic activity is lost. A few proteins are also denatured if the temperature is reduced to below 10°C.

3.2.4 Denaturation of Proteins

On treatment with acids, alkalies, concentrated saline solutions, solvents, heat, and radiations, the pattern of charges carried by the ionizable side chains of a protein changes entirely from that under normal conditions, so that the compact structure is usually disrupted, a more random structure being formed, this process is termed denaturation.

Protein denaturation is any modification in conformation (secondary, tertiary or quaternary) not accompanied by the rupture of peptide bonds, involved in primary structure. Since the tertiary structure of an active globular protein is characterized by the majority of hydrophobic groups being hidden inside the molecule, disruption will bring these in contact with the aqueous solvent and the solubility of the protein will decrease considerably.

Certain proteins are already unfolded in their native state (casein monomers) which explains their stability towards certain denaturing agents including heat. The effects of protein denaturing are numerous but the following needs special attention:

1) decreased solubility, resulting from the unmasking of hydrophobic groups;
2) altered water-binding capacity;
3) loss of biological activity (e.g., enzymic or immunological);
4) increased intrinsic viscosity;
5) inability to crystallize.

SUMMARY

Genetic information is stored in cells as the base sequence of DNA. This can be replicated prior to cell division, so that each cell produced contains the same genetic information as the original cell. The information may be transcribed into RNA structure and then translated into protein structure in such a way that the amino acid sequence of each protein synthesized is determined by the base sequence of a section of DNA known as gene.

The chemical and acid-base properties of proteins are largely those of the side chains of the amino acid residues present. The solubility of a protein in an aqueous solvent is influenced by salt concentration, pH, organic solvent content, and temperature.

RECOMMENDATIONS

1. Stryer, L. (1988), *Biochemistry*, 3rd edn. (Chapters 2-5, 27-33), Freeman.
2. Voet, D. and Voet, J.G. (1990), *Biochemistry* (Chapters 7, 27-34), John Wiley.
3. Walker, R. (1983), *The Molecular Biology of Enzyme Synthesis*, John Wiley.
4. Zubay, G. (1988), *Biochemistry*, 2nd edn. (Chapters 7, 26-31), Macmillan.

4 Specificity of Enzymes

Enzymes are specific in their action. This simply means that a proteolytic enzyme will not hydrolyse the glycosyl or ester bonds, in a carbohydrate or a lipid, nor it will hydrolyse all peptide bonds in a protein. For example, both α-chymotrypsin and trypsin are proteolytic enzymes but α-chymotrypsin hydrolyses peptide bonds in which the carbonyl group of that bond is supplied by tyrosine, phenylalanine or tryptophan whereas trypsin hydrolyses peptide bonds in which the carbonyl group of the peptide bond belongs to arginine or lysine.

4.1 TYPES OF SPECIFICITY

Enzymes show the following varying degrees of specificity:

4.1.1 Bond Specificity

The enzyme does not discriminate among substrates, but exhibits specificity only toward the bond being split. Esterases and proteases exhibit bond specificity.

4.1.2 Group Specificity

The enzymes may act on several different, though closely related, substrates to catalyse a reaction involving a particular chemical group. An example of this kind of enzyme is alcohol dehydrogenase, which catalyses the oxidation of a variety of alcohols. Another is hexokinase, which assists the transfer of phosphate from ATP to several different hexose sugars. Trypsin is specific for peptide bonds on the carboxyl side of arginine and lysine.

4.1.3 Absolute Specificity

The enzyme attacks only one substrate and catalyses only a single reaction. Most enzymes fall into this category. Glucose kinase exhibits absolute specificity in exchanging phosphate only between ATP and glucose.

4.1.4 Stereochemical Specificity

Enzymes demonstrate complete stereospecificity in catalysis, and can distinguish between optical or geometric isomers. Enzymes almost always use one form of an enantiomeric pair, unless their specific function is to catalyse the isomerization of enantiomers. For example, L-amino acid oxidase mediates the oxidation of L-amino acids to oxo-acids; a separate enzyme, D-amino oxidase, is required for the corresponding oxidation of D-amino acids. Even greater specificity is shown by the fungal enzyme glucose oxidase, which catalyses the reaction:

No other naturally occurring sugar, including α-D-glucose and α-D-galactose, can be acted upon to any appreciable extent.

Enzyme-catalysed reactions may yield stereospecific products, even when the substrate possesses no asymmetric carbon atom. For example, the action of glycerol kinase on glycerol always results in the production of L-glycerol-3-phosphate.

$$\text{Glycerol} + \text{ATP} \rightleftharpoons \text{L-glycerol-3-phosphate} + \text{ADP}$$

Glycerol   L-glycerol-3-phosphate

No L-glycerol-1-phosphate is formed, even though the two -CH_2OH groups of glycerol are chemically identical.

In addition to being substrate specific, enzyme-catalysed reactions are product-specific too.

4.2 ROLE OF THE 3D STRUCTURE IN SPECIFICITY OF ENZYMES

It is not just the unique amino acid sequence on the basis of which the complete specificity of an enzyme can be accounted. As mentioned earlier, enzymes structure is often shared by a cofactor. When this is the case then catalytic activity of the enzyme is not only the function of the amino acid sequence, higher levels of enzyme structure also become important.

Enzymes, unlike non-biological catalysts, may be extremely specific in their catalytic behaviour leads to the question, how all enzymes, inspite of having the protein structure in general, are so different in terms of their catalytic action. This cannot be explained without exploiting the term 'active site' of an enzyme. In order to explain the term active site and its structure, Ogston (1948) pointed out that there are different points of interactions between enzyme and substrate. These interactions can have either a binding or a catalytic function. The binding sites link to specific groups in the substrate, ensuring that the enzyme and substrate molecules are held in a fixed orientation with respect to each other, with the reacting group or groups in the vicinity of catalytic sites. The binding and catalytic sites must be either amino acid residues or cofactors, the latter being themselves bound to amino acid side chains. The region of the enzyme's three-dimensional structure which contains the substrate-binding sites and the catalytic sites is termed the active site or active centre of the enzyme.

The size of an active site varies because the number of interacting positions between the substrate and enzyme on binding is often a function of the substrate size. In a large polymer, such as a starch or protein molecule, there may be a very large number of interacting positions. Granted this, an active site still comprises only a small proportion of the total volume of the enzyme and is usually at or near the surface, since it must be accessible to substrate molecules. Thus an enzyme molecule is, when compared to its substrate, often larger in size by a factor of several orders of magnitude. For example, glucose oxidase is 150 Kdal while glucose is 180 dal. This strongly suggests that a large proportion of the peptide chain of the enzyme cannot be in direct contact with the substrate. Then how does this remaining portion of the protein molecule function?

The answer to this lies in the fact, that the groups involved in the active site are usually not in proximity in terms of the primary sequence of the protein chain. The groups are well separated in terms of primary sequence but must be brought in proximity through the turns and twists of the tertiary structure. Therefore, a large part of the remainder of the peptide chain would be involved in maintaining the conformational features of the active site, i.e. in keeping all these groups properly positioned in the active site. A change in the relative position of one of the groups by as little as a fraction of one angstrom unit may be sufficient to cause loss of activity.

From the above discussion, it must be clear that the active site of an enzyme is more than the sum of its parts. In order to understand how an enzyme works, it is necessary to know not only the identities of the amino acid residues which make up the substrate-binding sites and catalytic sites, but also the environment in which each site exists and the arrangement in space of the sites with respect to each other.

4.2.1 Stereospecific Binding of an Active Site

A substance must meet the very rigid restrictions of structure as imposed by the enzyme before it is bound by the binding site. There are distinct recognition areas within the binding locus. Consider, for example, the molecule of ethanol ($CH_3 \cdot CH_2 \cdot OH$). Ethanol has no asymmetric carbon atom, yet alcohol dehydrogenase must recognize a difference between the two methylene hydrogens, since it stereospecifically removes the same hydrogen atoms (one methylene hydrogen and the other hydroxyl hydrogen) from the molecule to produce acetaldehyde. It does this repeatedly when the reaction is reversed several times in each direction. The same hydrogen of the methylene group is always removed or added. To explain this, it has been assumed that the enzyme must bind simultaneously to more than one point of the molecule.

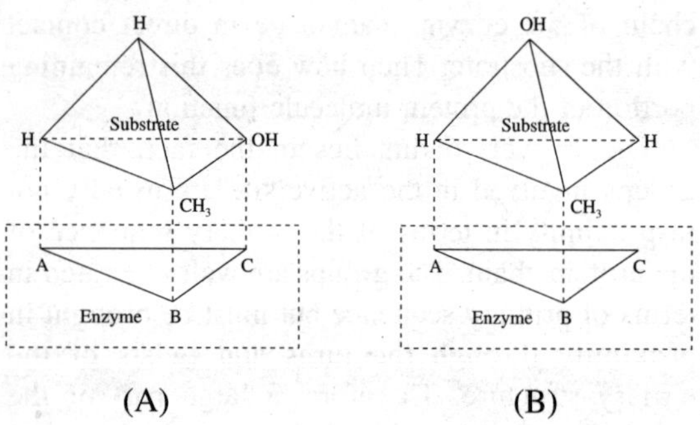

Fig. 4.1 *Possible modes of orientation of ethanol on the surface of alcohol dehydrogenase.*

Thus, when two substituents (e.g. the methyl and hydroxyl groups of ethanol) are attached to the enzyme surface at positions B and C, the position of third group is fixed. Only 'A' leads to correct alignment for transformation to product. Two equal substituents in a symmetrical molecule are differentiated by asymmetric binding to the enzyme.

4.2.2 The Fischer 'Lock-and-Key' Hypothesis

As early as 1980, Fisher suggested that enzyme specificity implies the presence of complementary structural features between enzyme and substrate. A substrate may fit into its complementary site on the enzyme as a key fits into the lock. According to this hypothesis, all structures remain fixed throughout the binding process.

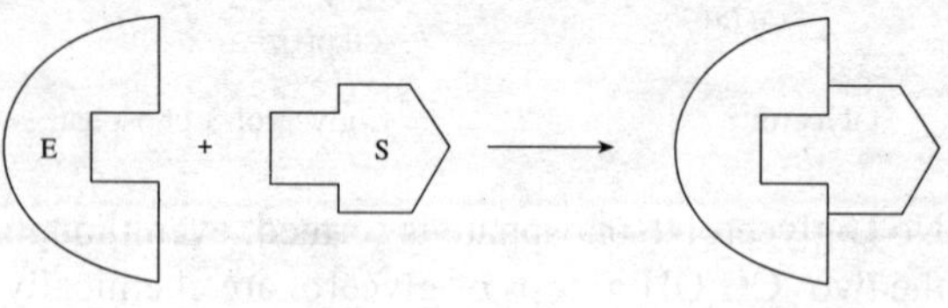

Fig. 4.2 *Representation of 'Lock-and-Key' hypothesis.*

4.2.3 The Koshland 'Induced-fit' Hypothesis

The lock-and-key hypothesis explains many features of enzyme specificity, but takes no account of the known flexibility of proteins. X-ray diffraction analysis and data from several forms of spectroscopy, including nuclear magnetic resonance (NMR), have revealed differences in structure between free and substrate-bound enzymes.

Thus, the binding of a substrate to an enzyme may bring about a conformational change, i.e. a change in three-dimensional structure but not in primary structure. The bonds that form between a substrate and binding sites of enzyme may replace previously existing linkages between each binding site and neighbouring groups on the enzyme. Also, the presence of a substrate at the active site may exclude water molecules and thus make the region more non-polar. Both of these factors may be responsible for some degree of change in tertiary structure taking place.

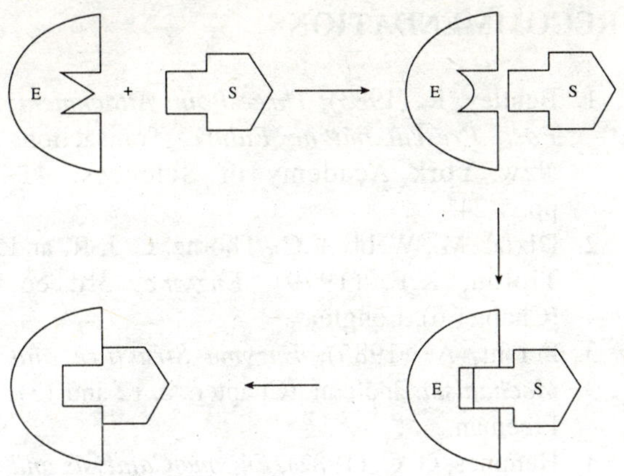

ES-Complex

Fig. 4.3 *A model depicting the 'Induced-fit' hypothesis.*

Koshland, in his induced-fit hypothesis, suggested that the structure of a substrate may be complementary to that of the active site in the enzyme-substrate complex, but not in the free enzyme. A conformational change takes place in the enzyme during the binding of substrate which results in the required matching of structures. The induced-fit hypothesis essentially requires the active site to be floppy and the substrate to be rigid, allowing the enzyme to wrap itself around the substrate, in this way bringing together the corresponding catalytic sites. In some respects, the relationship between a substrate and an active site is similar to that between a hand and a woolen glove; in each interaction the structure of one component (substrate or hand) remains fixed and the shape of the second component (active site or glove) changes to become complementary to that of the first.

Such a mechanism can help to achieve a high degree of specificity for the enzyme. In the lock-and-key mechanism the active site is always structurally intact with the catalytic sites aligned and freely accessible. Thus a suitable reacting group, whether part of an appropriately bound substrate or not, can come into contact with the region of catalytic activity and some degree of reaction take place. In the induced-fit mechanism, on the other hand, different catalytic components might be separated by a considerable margin in

the free enzyme minimizing the risk of a chance of collision of a reactive group with both of them. It is also possible that access to the catalytic groups of the free enzyme might be blocked. Only when a binding group of the substrate is recognized by the corresponding site of the enzyme and the binding process proceeds does the conformational change take place which results in all the relevant groups in substrate and enzyme coming together. Of course, a similar binding group in a substance other than the substrate might trigger off a conformational change but, in general, this would not result in catalytic groups being brought in the vicinity of an appropriate reacting group, so no reaction would take place. This would be termed non-productive binding.

4.2.4 Strain or Transition-state Stabilization Hypothesis

Although the lock-and-key and induced-fit models can explain enzyme specificity, neither of the two suggests any direct mechanism by which the catalysed reaction may be driven forward. Substrate-binding often involves the expenditure of a considerable amount of energy and although it serves a very useful purpose of bringing reacting and catalytic groups together, further energy must be supplied before the reaction can proceed. If the binding energy is used to distort the substrate in such a way as to facilitate the subsequent reaction then less energy would be required for the reaction to take place.

If the structure of the active site is rigid, the substrate must be distorted slightly in order to bind with the enzyme. This distortion might result in the stretching, and weakening, of a bond which is subsequently to be cleaved, thus assisting the forward reaction.

An alternative, and possibly more likely, mechanism for driving the reaction forward is transition-state stabilization. This assumes that the substrate is bound in an undistorted form but the enzyme-substrate complex possesses various unfavourable interactions. These tend to distort the substrate in such a way as to favour the following reaction sequence:

enzyme-substrate complex———→transition-state———→products

As the reaction proceeds, the unfavourable interactions diminish and are absent from the transition-state.

SUMMARY

Enzymes exhibit chemical and stereochemical specificity with respect both to substrates and products. Such specificity requires at least three different points of interaction between enzyme and substrate. Each substrate is bound to the enzyme at specific sites to form an enzyme-substrate complex in which reacting groups are held in close proximity to each other and to catalytic sites. That region of the enzyme's three-dimensional structure which contains the substrate-binding sites and the catalytic sites is termed the active site.

According to the Fischer lock-and-key hypothesis, the active site has rigid structural features which are complementary to those of each substrate. In contrast, the Koshland induced-fit hypothesis suggests that at least some active sites are flexible, possessing a structure complementary to that of a substrate only when the latter is bound to the enzyme. These models can explain some aspects of enzyme specificity, but do not suggest any mechanism for driving forward the enzyme-catalysed reaction. The concept of strain or the transition-state stabilization explains how distortion of the substrate during or after binding can facilitate the subsequent reaction. This is not necessarily inconsistent with an induced-fit mechanism.

The most important factor in determining whether an enzyme will act on a particular substrate to produce a product appears to be the stability of the enzyme-bound transition-state which would have to be formed.

RECOMMENDATIONS

1. Bentley, R. (1983), *Three-Point Attachment: Past, Present but no Future*, Transactions New York Academy of Sciences, 41, pp. 5-24.
2. Dixon, M., Webb, E.C., Thorne, C. J. R. and Tipton, K.F. (1979), *Enzyme*, 3rd ed. (Chapter 6), Longman.
3. Fersht, A. (1985), *Enzyme Structure and Mechanism*, 2nd edn. (Chapters 2, 12 and 13), Freeman.
4. Hammes, G. G. (1982), *Enzyme Catalysis and Regulation* (Chapter 6), Academic Press.
5. Mackenzie, N.E., Malthouse, J.P.G. and Scott, A. I. (1984), *Studying Enzyme Mechanism by ^{13}C NMR*, Science, 225, pp. 883-9.
6. Stryer, L. (1988), *Biochemistry*, 3rd edn. (Chapter 8), Freeman.

5 Introduction to Catalysis and Kinetics

5.1 FACTORS WHICH AFFECT THE RATE OF CHEMICAL REACTIONS

5.1.1 The Collision Theory

Molecules can react only if they come into contact with each other. Therefore, any factor which increases the rate of collisions, e.g. increased concentration of the reactants or increased temperature, will increase the reaction rate.

However, all colliding molecules do not react. They can be partly explained by steric reasons, for not all collisions will result in the appropriate groups of molecules coming into contact, particularly if the reactants are complex. A further and more important reason is that not all colliding molecules possess sufficient energy between them to undergo a reaction.

5.1.2 Activation Energy and the Transition-state Theory

Not all molecules of the same type will possess the same amount of energy, taking all forms of energy into account. The energy of an individual molecule will depend, for example, on what collisions that molecule has recently been involved in. In order for a reaction to take place, colliding molecules must have sufficient energy to overcome a potential-barrier known as the energy of activation. This is true even of energetically favourable reactions, i.e. those which can proceed spontaneously with liberation of free energy.

The best explanation for the requirement of activation energy in a chemical reaction is the transition-state theory. This postulates that every chemical reaction proceeds via the formation of an unstable intermediate between reactants and products.

Consider the hydrolysis of an ester :

$$RC\underset{OR}{\overset{O}{\diagup}} + H_2O \longrightarrow RC\underset{OH}{\overset{O}{\diagup}} + ROH$$

It is believed that the reaction mechanism involves the addition of water to the ester to form a transition-state compound having regions of positive and negative charge which cannot be stabilized.

An unstable compound, by definition, will break down to give a more stable one and so possess more free energy than the stable compound. A free energy profile of such a reaction involving unstable transition-state is illustrated as :

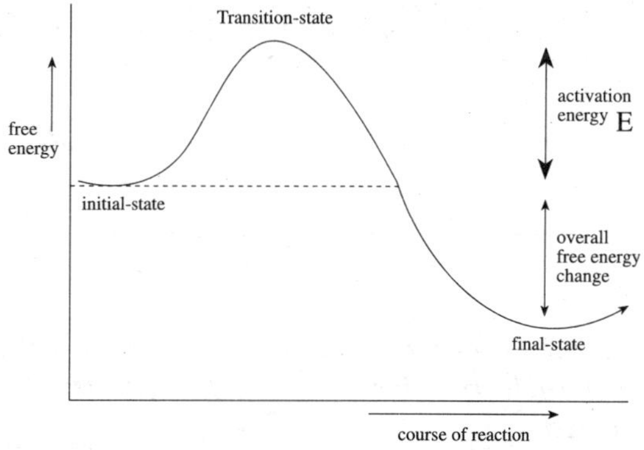

Fig. 5.1 *Free energy changes for an energetically favourable reaction proceeding via the formation of a transition state.*

The activation energy is, therefore, the energy needed to form the transition-state from the reactants. The transition-state is unstable and will very quickly break down to form the products (or break to reactants) but no products can be formed from reactants unless the transition-state has been formed. The free energy of activation thus acts as a potential-barrier to the reaction taking place.

5.1.3 Catalysis

A catalyst accelerates a chemical reaction without changing its extent and can be removed unchanged from amongst the end-products of the reaction. It has no overall thermodynamic effect; the amount of free energy liberated or taken up when a reaction has been completed will be the same whether a catalyst is present or not.

In most cases, a catalyst acts by reducing the energy of activation. The catalyst, or part of it, combines with reactants to form a different transition-state from that involved in the uncatalysed reaction; one which is more stable and therefore, of lower energy.

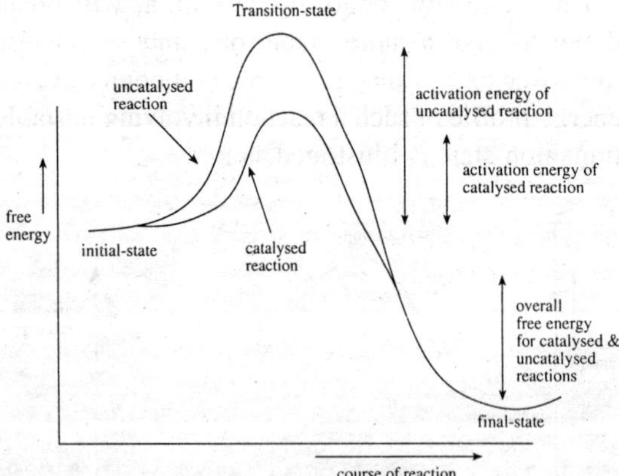

Fig. 5.2 *Free energy changes for an energetically favourable reaction, showing the effects of a catalyst.*

Note that the initial and final states are at the same free energy levels for the catalysed and uncatalysed reactions, and the overall free energy change as the reaction proceeds is also the same.

An uncatalysed reaction, therefore, may proceed as follows:

$$A+B \xrightarrow{\text{Very slow}} A\text{-}\text{-}\text{-}B \xrightarrow{\text{Very fast}} \text{Product}$$

The same reaction in presence of a catalyst C may take place like this :

$$A+B+C \xrightarrow{\text{Very slow}} A\text{-}\text{-}\text{-}B\text{-}\text{-}\text{-}C \xrightarrow{\text{Very fast}} \text{Product} + C$$

Catalysts are often acids or bases: acids stabilize the transition-state by donating a proton, bases by accepting a proton. The hydrolysis of an ester, which was considered earlier, may be catalysed by a base ($X\text{---}O^-$) as follows :

Other common forms of catalysis include covalent catalysis, where the transition-state is stabilized by changes involving covalent bonds, and metal ion catalysis, where the transition-state is stabilized by electrostatic interactions with a metal ion.

5.2 KINETICS OF UNCATALYSED CHEMICAL REACTIONS

5.2.1 The Law of Mass Action and the Order of Reaction

Kinetics is the study of reaction rates and the factors influencing them. It is not concerned with the chemical nature of the changes taking place.

All kinetic work is based on the Law of Mass Action. This states that 'the rate of a reaction is proportional to the product of the activities of each reactant, each activity being raised to the power of the number of molecules of that reactant taking part, as indicated by the reaction equation'.

$$aA + bB \longrightarrow \text{Products}$$

The rate is proportional to activity (activity of A)^a x (activity of B)^b.

For practical purposes, the term activity is usually replaced by concentration.

The concept of the order of reaction was developed from the Law of Mass Action. A first-order reaction is one which proceeds at a rate proportional to the concentration of one reactant. Thus, for a first-order reaction 'A $\rightarrow$ P' taking place at a constant temperature and pressure in a dilute solution, reaction rate at any time 't' is given by:

$$v = -\frac{d[A]}{dt} = +\frac{d[P]}{dt} = k[A]$$

Where v = reaction rate at time t
     k = rate constant
    [A] = concentration of reactant A at time t
    [P] = concentration of product P at time t

$$-\frac{d[A]}{dt} = \text{rate of decrease of } [A]$$

$$+\frac{d[P]}{dt} = \text{rate of increase of } [P]$$

A second-order reaction is one which proceeds at a rate proportional to the concentrations of two reactants or to the square of the concentration of a single reactant. So, for a second-order reaction 'A + B $\longrightarrow$ P', taking place at a constant temperature and pressure in a dilute solution,

$$v = -\frac{d[A]}{dt} = -\frac{d[B]}{dt} = +\frac{d[P]}{dt} = k[A][B]$$

the terms having the same general meanings as above. Similarly, for a second-order reaction of the form '2A $\longrightarrow$ P',

$$v = -\frac{d[A]}{dt} = +\frac{d[P]}{dt} = k[A]^2$$

A two-reactant reaction can, under certain circumstances, be regarded as a pseudo single-reactant one. For instance, reactions taking place in an aqueous medium and involving water and one other reactant are usually considered pseudo single-reactant since water will be in vast excess and its concentration will not change significantly during the course of the reaction.

It should also be noted that a zero-order reaction is possible. This is one whose rate is independent of the concentration of any of the reactants.

5.2.2 The Use of Initial Velocity

For any reaction, starting with reactants only and measuring the appearance of product with time, a graph of the form illustrated below is obtained.

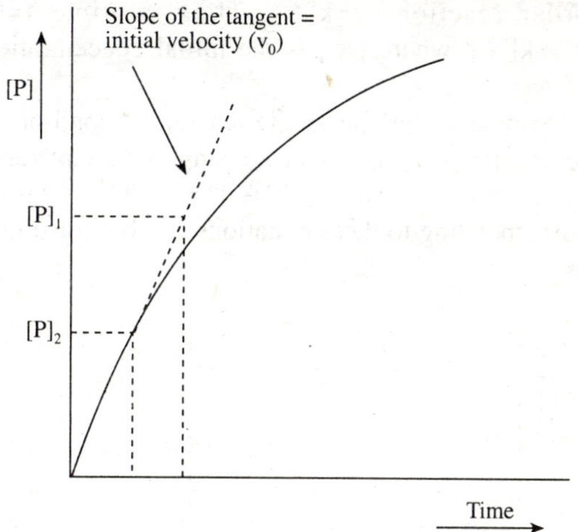

Fig. 5.3 *Graph of product concentration against time for a chemical reaction.*

The rate of the reaction at any time 't' is the slope of the curve at that point. This may be constant for a little time at the start of the reaction and then decreases with the decreasing concentration of the reactant(s) as the reaction proceeds, finally falling to zero. At this point either all the reactants have been converted into products or, more commonly, a chemical equilibrium has been set up, with the rate of the forward reaction now being equal to the rate of the back reaction.

The initial velocity (v_0) of the reaction is the reaction rate at t=0 and may be determined by drawing a tangent to the graph, as shown in figure. From this tangent,

$$v_0 = \frac{[P]_2 - [P]_1}{t_2 - t_1}$$

The units for v_0 are those used for product concentration, divided by those used for time.

The importance of initial velocity is that it is a kinetic parameter determined for the reaction in a situation which can be easily specified. The concentration of each reactant is known from the amount actually added; this would not be true at any other point during the reaction. Also, since there are no products present at t=0, no back reaction will be taking place.

The initial velocity depends on the initial concentration of the reactants. As discussed in the previous section that, for a single-reactant first-order reaction, $v=k[A]$. Thus, at time t=0, $v_0 = k[A_0]$, where $[A_0]$ is the initial concentration of A.

Similarly, for the single-reactant second-order reaction, $v_0 = k[A_0]^2$ and for a single-reactant zero-order reaction, $v_0 = k[A_0]^0 = k$. The graphs corresponding to these equations can be illustrated as :

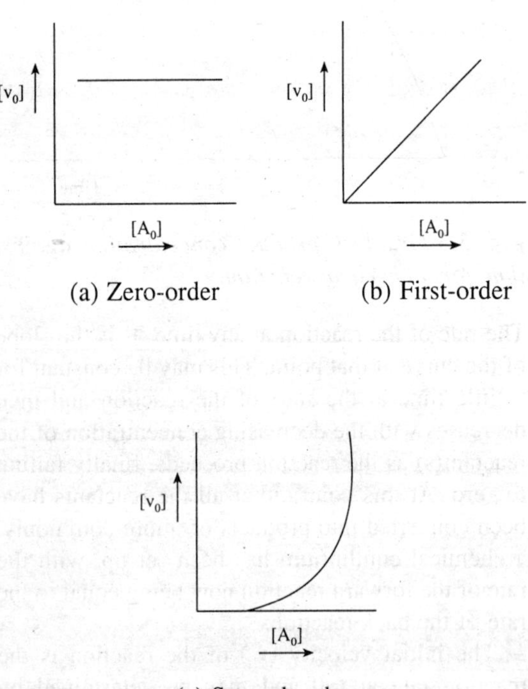

Fig. 5.4 *Graph of initial velocity against initial reactant concentration for a single-reactant reaction.*

5.3   KINETICS OF ENZYME-CATALYSED REACTIONS

Since the rate of acid-hydrolysis of sucrose is proportional to the sucrose concentration at constant acid concentration, therefore, we can say that this reaction is first order with respect to sucrose.

When the identical reaction, but catalysed by the enzyme invertase, is similarly carried out, different results are obtained. At low sucrose concentrations the reaction is again first-order with respect to the substrate, but at higher concentrations it becomes zero-order.

This is generally found to be true for all single-substrate enzyme catalysed reactions and for all multi-substrate reactions where the concentration of all the substrates but one is kept constant. A graph of initial velocity (v_0) against initial substrate concentration ($[S_0]$) at constant total enzyme concentration ($[E_0]$) is found to be a rectangular hyperbola.

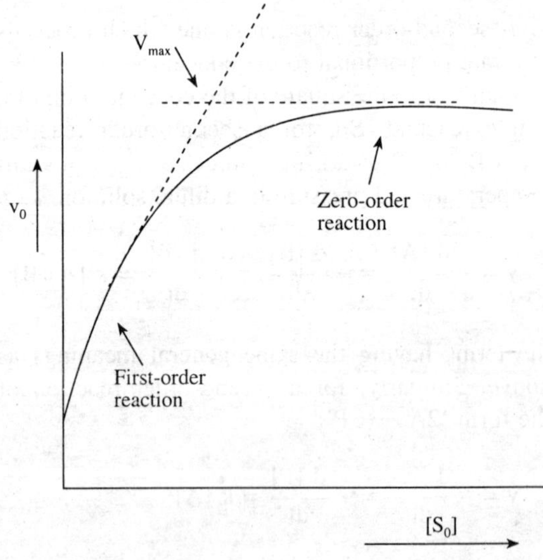

Fig. 5.5 *Graph of initial velocity against initial substrate concentration at constant total enzyme concentration for a single substrate enzyme-catalysed reaction.*

Such a graph has a general equation

$$v_0 = \frac{a[S_0]}{[S_0] + b}$$

where 'a' and 'b' are constants. The constant 'a' represents the maximum value of v_0 (called V_{max}) and 'b' is the value of $[S_0]$ where $v_0 = 1/2\ V_{max}$.

Although some enzymes do not give hyperbolic graphs, the attainment of a maximum initial velocity with increasing substrate concentration at constant total enzyme concentration is characteristic of all enzymes.

The explanation for this feature was first given by Brown, with particular reference to the hydrolysis of sucrose but subsequently found to be of general application. Enzyme and substrate combine to form an enzyme-substrate complex, which undergoes a further reaction to breakdown to enzyme and products:

$$E + S \xrightarrow{\text{rate constant } k_1} ES \xrightarrow{\text{rate constant } k_2} E + P$$

The overall rate of reaction (the rate of formation of P) must be limited by the amount of enzyme available and by the rate of breakdown of the enzyme-substrate complex. If the substrate concentration is sufficiently high, it will saturate the enzyme, i.e. force an immediate reaction with each available enzyme molecule to form an enzyme-substrate complex. Under these conditions, therefore, there will be no free enzyme present and the concentration of enzyme-substrate complex ($[ES]$) will be the total enzyme concentration present ($[E_0]$), making the overall rate of reaction $k_2\ [E_0]$ (from the Law of Mass Action). This is independent of substrate concentration and so cannot be increased by using still higher substrate concentrations. It is, therefore, the maximum initial velocity possible at this enzyme concentration. Hence,

$$V_{max} = k_2\ [E_0]$$

In contrast, at very low substrate concentrations the enzyme will be far from saturation and the overall rate of reaction will be limited by the rate at which enzyme and substrate molecules react to form an enzyme-substrate complex. At constant enzyme concentration this will be proportional to the substrate concentration and, therefore, a first-order reaction will result.

At intermediate substrate concentrations, the enzyme will be partially saturated with substrate and the order of reaction will be somewhere between zero-order and first-order.

The terms 'saturated' and 'partially saturated', refer to the population of enzyme molecules and not to each individual molecule. An enzyme molecule which binds a single substrate molecule cannot itself be partially saturated at any given moment but a population of such molecules can be; some being substrate-bound and some free.

5.3.1 Methods Used for Investigating the Kinetics of Enzyme-Catalysed Reactions

5.3.1.1 Initial velocity studies

Initial velocity studies have been particularly useful for investigating the kinetics of enzyme-catalysed reactions because restrictions of investigations to v_0 determinations give the best chance of avoiding errors caused by loss of enzyme activity with time.

The initial velocity of a reaction is usually determined from a graph of product concentration against time, alternatively, a direct instrumental reading known to be proportional to product concentration (e.g. absorbance units) may be plotted against time. The rate of reaction could also be determined from the disappearance of substrate as the reaction proceeds but in general, it is considered a better technique to measure the appearance of something from an initial value of nothing, rather than the disappearance of something from an initial large value.

The experiments are performed at constant temperature and pH using a method which enables the course of the reaction to be monitored continuously and preferably automatically. The actual technique used depends on the reaction being investigated.

If there is a difference between substrate and product in the absorbance of light of a particular wavelength, spectrophotometric or colorimetric techniques may be used. The molar extinction coefficient of a substance, if known, enables the actual concentration of that substance to be calculated from an absorbance reading:

absorbance = molar extinction coefficient x concentration (mol L) x cell path length (cm)

Reactions where gases are evolved or taken up as the reaction proceeds can be investigated by manometric techniques; the reaction is performed in an airtight vessel coupled to a manometer, which records changes in volume or pressure. Electrochemical methods can be applied wherever there are differences in electrical properties between substrate and product.

5.3.1.2 Rapid-reaction techniques

To investigate the formation and breakdown of the enzyme-substrate complex or complexes, it is necessary to use techniques capable of detecting changes taking place over time scales of the order of magnitude of 1 millisecond. Of particular importance, is a detailed kinetic study of the changes taking place over the first fraction of a second of the reaction (termed transient kinetics). This is usually performed using rapid mixing techniques of the continuous-flow or stopped-flow variety, the reaction being monitored by some suitable detector coupled to an oscilloscope and usually to a computer.

With continuous-flow systems, streams of enzyme and substrate converge are pumped together at a fixed speed down capillary-bore tubing to ensure rapid mixing and elimination of dead-space. From a knowledge of the dimensions of the tube and the flow rate, the time taken from mixing to the arrival at the detector can be calculated. By altering the flow rate, the time of reaction can be changed.

With stopped-flow techniques, solutions of enzymes and substrate are rapidly injected together into an observation chamber and the course of reaction monitored continuously.

The main limitation of both these techniques is the time taken to mix enzyme and substrate. This problem does not apply to the alternative approach, the investigation of relaxation kinetics. Here the enzyme and substrate are mixed and the system allowed to come to equilibrium, then the position of the equilibrium is rapidly altered by a sudden temperature or pressure change, and the approach to the new position of equilibrium monitored.

5.3.2 The Nature of Enzyme Catalysis

Enzymes behave like any other catalyst in forming with the reactants a transition state of lower free energy than that which would be found in the uncatalysed reaction. For a single-substrate reaction the enzyme initially binds the substrate at a specific binding-site to form a relatively stable enzyme-substrate complex. This process takes place via the formation of an unstable transition-state. In the enzyme-substrate complex, the reacting groups are held in close proximity to each other and to the catalytic-site of the enzyme. The catalysed reaction can now take place, via the formation of another unstable transition-state, to give the product. However, at this point the product may still bind to the enzyme, in which case another relatively stable reaction intermediate, the enzyme-product complex, would exist before the free product was liberated. The free energy profile of a reaction of this type can be represented as:

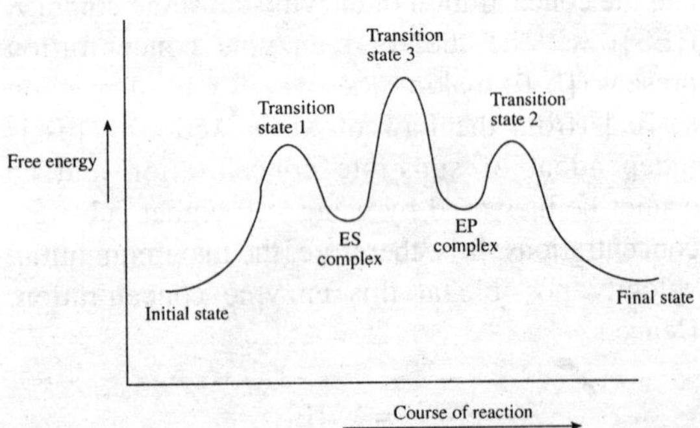

Fig. 5.6 *The free energy profile of an enzyme-catalysed reaction involving the formation of an enzyme-substrate and an enzyme-product complex.*

For a reaction of the form:

or

$$E + S \xrightarrow{k_1} ES \xrightarrow{k_2} E + P$$

$$E + S \xrightarrow{k_1} ES \xrightarrow{k_2} EP \xrightarrow{k_3} E + P$$

Where $ES \longrightarrow EP$ is the rate-limiting step

The overall rate of reaction is given by k_2 [ES] and the best estimate of k_2 comes from V_{max}, since $V_{max} = k_2 [E_0]$.

SUMMARY

Biochemical processes can only proceed spontaneously, in such a direction that the free energy of the system, i.e. the energy that can be used to perform work, decreases. However, even energetically favourable chemical reactions have to overcome a potential-barrier, known as the activation energy, before the reaction can take place. This is explained by the need to form unstable transition-states. Catalysts, including enzymes, act by allowing the formation of different, more stable, transition-states and, thus, reduce the activation energy. The position of chemical equilibrium is unchanged but is reached much faster than in the equivalent uncatalysed reaction.

The initial velocity of enzyme-catalysed reactions has a limiting value at each total enzyme concentration; this occurs when the enzyme is saturated with substrate. Enzymes react with substrate to form enzyme-substrate complexes; these are quite distinct from the transition-states which also occur as part of the process of enzyme catalysis.

RECOMMENDATIONS

1. Dawes, E.A. (1980), *Quantitative Problems in Biochemistry*, 6th edn. (Chapters 3-5), Longmam.
2. Harold, F.M. (1986), *The Vital Force : A Study of Bioenergetics* (Chapters 1-2), Freeman.
3. Morris, J.G. (1974), *A Biologist's Physical Chemistry*, 2nd edn. (Chapters 7-10), Edward Arnold.

6 Kinetics of Single-Substrate Enzyme Catalysed Reactions

6.1 THE TIME COURSE OF AN ENZYME REACTION

Generally we can recognize three phases for the overall time course of an enzyme reaction. These are :

Phase I   : The transient initiation phase
             a few seconds' duration.
Phase II   : A steady-state phase
             the initial reaction rate.
Phase III : A non-linear phase
             the main reaction up to completion.

6.1.1 Phase-I: Reaction Initiation

An enzyme-catalysed reaction is initiated by the formation of an enzyme substrate complex which subsequently breaks down to product and regenerates the original enzyme.

$$E + S \rightleftharpoons ES \longrightarrow P + E$$

The rate limiting step is assumed to be the breakdown of enzyme substrate complex.

$$ES \longrightarrow P + E$$

During the initiation of the reaction, the free enzyme concentration (usually only a fraction of the substrate concentration) drops to almost zero as a steady state is established between the enzyme and the substrate.

6.1.2 Phase-II: Initial Reaction

All the reactants are in dynamic equilibrium and therefore operate at maximum efficiency under a near-ideal steady-state system.

6.1.2.1 Reaction progress curves

If we follow the transformation of substrate with time for enzyme E under constant physical conditions, a curve, similar to that shown in the figure, is usually obtained.

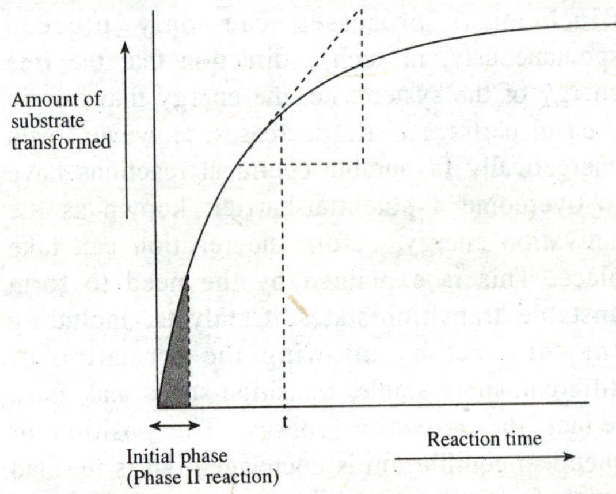

Fig. 6.1 *Progress curve for a typical enzyme reaction. A tangent drawn at time 't' to the reaction curve gives the instantaneous reaction velocity. The phase II part of the reaction is shown as the shaded area.*

The figure shows that when the reaction is proceeding at its fastest, the reaction curve is composed of an initial linear portion which gives way to a curve at longer reaction times. This linear portion is the initial reaction rate and thus the phase II of the reaction.

6.1.2.1a Effect of substrate concentration

For constant enzyme concentration, as the reaction proceeds, available substrate becomes a contributing limiting factor in the reaction rate. It is thus important to look at the effect of substrate concentration on the initial rate of the reaction.

Measurement of the initial velocity as a function of initial substrate concentration results in a curve as shown in figure 6.2. The shape of this curve is very significant and can give us a great deal of useful information. To begin with, it is in the form of a rectangular hyperbola.

(i) Reaction at high substrate concentration

At $v = V_{max}$, reaction rate is independent of substrate concentration which tells us that the reaction occurring is of zero order. In describing the phase I reaction, it was assumed that the overall rate limiting step was the breakdown of ES to $E + P$.

$$E + S \rightleftharpoons ES \longrightarrow P + E$$

By rewriting this equation and putting the separate velocity constants,

$$E + S \underset{k_{-1}}{\overset{k_1}{\rightleftharpoons}} ES \overset{k_2}{\longrightarrow} P + E$$

It can be seen that:

$$v = V_{max} = k_2 [ES]$$

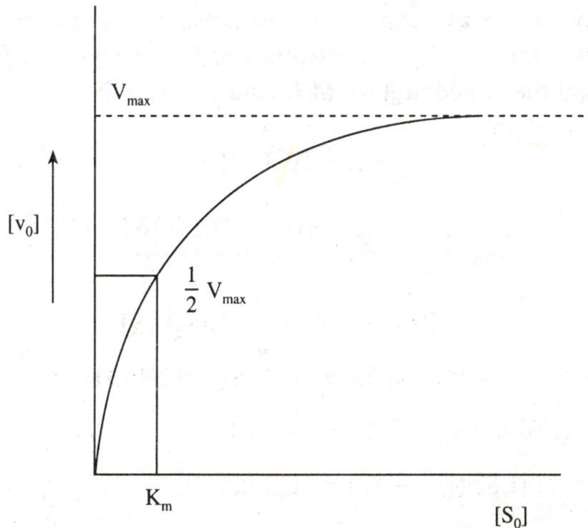

Fig. 6.2 *Graph of initial velocity as a function of substrate concentration.*

At any instant, in a reaction in which an enzyme is participating, the enzyme is present in both the free state E and combined as the enzyme substrate complex, therefore if the total molar enzyme concentration at the initiation of the reaction is E_0, then:

$$[E_0] = [E] + [ES]$$

Usually in an enzyme reaction and specially at high substrate concentration $[S] > E_0$, therefore the equilibrium governing the formation of the enzyme substrate complex:

$$E + S \underset{k_{-1}}{\overset{k_1}{\rightleftharpoons}} ES$$

is completely in favour of complex formation which means that [E] is virtually zero, therefore equation $[E_0] = [E] + [ES]$ reduces to $[E] = [ES]$, and equation $v = k_2 [ES]$ becomes $v = V_{max} = k_2 [E_0]$.

(ii) Reaction at low substrate concentration

At low substrate concentrations figure 6.2 indicates that part of the curve is virtually linear, hence initial velocity is directly proportional to substrate concentration, that is,

$$\upsilon = k' [S]$$

This means that reaction is now of the first order type. On the phase I , this means that [ES] is low particularly if k_1/k_{-1} is low.

(iii) Reaction at intermediate substrate concentration

What we require to know, is how the initial velocity is affected by substrate concentrations intermediate between these two extremes in which the first order reaction gradually reduces to a zero order reaction. Since we are considering only the initial reaction rates of substrate reacting with relatively low molar enzyme concentrations, the following assumptions must be valid:

$$(1)\ [S] > [E_0]\ \text{and}\ (2)\ [P] = 0$$

Using the simple phase I reaction, a further assumption can be made. Either the rate limiting step is the rate at which ES breaks down:

$$ES \xrightarrow{k_2} P + E \quad (\text{in which case } k_2 < k_{-1})$$

or $\dfrac{d[ES]}{dt} = 0$ that is [ES] is constant

Assumption (1) formed the basis on which the original Michaelis-Menten theory of enzyme kinetics was established. It was challenged by Haldane as being unnecessarily restrictive. Briggs and Haldane argued that since the concentration of enzyme, and thus enzyme-substrate complex, is usually very small compared with the substrate concentration, therefore the rate of change of [ES] would be negligible compared to the rate of change of [P] over the initial period of the reaction except during the very brief period of the beginning, while ES was first being formed. In the absence of the product the concentration of ES would be determined by the total enzyme concentration which remains constant throughout and by the substrate concentration, which changes by a negligible amount, as a percentage of its initial value, over the period of interest. So, once this complex has been produced it would be maintained in a steady-state, i.e. it would be broken down as fast as it was being formed, [ES] remaining constant.

The Henri and Michaelis-Menten equation

Let us consider a single-substrate enzyme-catalysed reaction where there is just one substrate-binding per enzyme. The simplest general equation for such a reaction would be:

$$E + S \underset{k_{-1}}{\overset{k_1}{\rightleftharpoons}} ES \underset{k_{-2}}{\overset{k_2}{\rightleftharpoons}} P + E$$

If investigations are restricted to the initial period of the reaction, the product concentration is negligible and the formation of ES from product can be ignored. Under these conditions, therefore, the reaction simplifies still further to:

$$E + S \underset{k_{-1}}{\overset{k_1}{\rightleftharpoons}} ES \xrightarrow{k_2} P + E$$

The rate of formation of ES at any time 't' (within the initial period when the product concentration is negligible) = k_1 [E][S], where [E] is the concentration of free enzyme and [S] the concentration of free substrate at time 't'.

Also at time 't', the rate of breakdown of ES back to E and S = k_{-1} [ES], where [ES] is the concentration of enzyme-substrate complex at this time.

The Michaelis-Menten assumes that an equilibrium between enzyme, substrate and enzyme-substrate complex sets up instantly and is maintained, the breakdown of enzyme-substrate complex to products being too slow to disturb this equilibrium. Using this assumption, therefore:

$$k_1 = [E][S] = k_{-1}\ [ES]$$

When constants are separated from variables, we get:

$$\frac{[E]\ [S]}{[ES]} = \frac{k_1}{k_{-1}} = K_s$$

Where K_s is the dissociation constant of ES. The total concentration of enzyme present $[E_0]$ must be the sum of the concentration of free enzyme [E] and the concentration of bound enzyme [ES]. Therefore,

$$[E] = [E_0] - [ES]$$

and

$$K_s = \frac{([E_0] - [ES])\ [S]}{[ES]}$$

$$[ES]\ K_s = ([E_0] - [ES])\ [S]$$

$$[ES]\ K_s = [E_0]\ [S] - [ES]\ [S]$$

$$[ES]\ [S] + [ES]\ K_s = [E_0]\ [S]$$

$$[ES]\ ([S] + K_s) = [E_0]\ [S]$$

$$[ES] = \frac{[E_0]\ [S]}{([S] + K_s)}$$

[ES] governs the rate of formation of products according to the relationship:

$$v_0 = k_2 \, [ES]$$

If we substitute the expression for [ES] derived above, we obtain:

$$v_0 = \frac{[E_0] \, [S]}{[S] + K_s}$$

Moreover, we know that when the substrate concentration is very high, all the enzyme is present as the enzyme-substrate complex and the limiting initial velocity V_{max} is reached. Under these conditions:

$$V_{max} = k_2 \, [ES]$$

Therefore, we can substitute V_{max} for $k_2 \, [E_0]$ in the expression for v_0 and get:

$$v_0 = \frac{V_{max} \, [S]}{[S] + K_s}$$

If we make the assumption that the initial substrate concentration $[S_0]$ is very much greater than the initial enzyme concentration $[E_0]$ then the formation of the enzyme-substrate complex will result in an insignificant change in free substrate concentration. Hence, in the expression for v_0 derived above, we can substitute $[S_0]$ for [S], giving:

$$v_0 = \frac{V_{max} \, [S_0]}{[S_0] + K_s}$$

Briggs-Haldane modification of the Michaelis-Menten equation

If we again consider a single-substrate single-biding-site reaction:

$$E + S \underset{k_{-1}}{\overset{k_1}{\rightleftharpoons}} ES \xrightarrow{k_2} P + E$$

the rate of formation of ES at any time $t = k_1 \, [E] \, [S]$

The rate of breakdown of ES at this time $t = k_{-1} \, [E] \, [S] + k_2 \, [ES]$, since ES can be broken down to form products or reform reactants.

Using the steady-state assumption:

$$k_1 \, [E] \, [S] = k_{-1} \, [E] \, [S] + k_2 \, [ES]$$
$$k_1 \, [E] \, [S] = [ES] \, (k_{-1} + k_2)$$

Separating the constants from the variables:

$$\frac{[E] \, [S]}{[ES]} = \frac{k_{-1} + k_2}{k_1} = K_m$$

where K_m is another constant. Substituting $[E] = [E_0] - [E]$,

$$K_m = \frac{([E_0] - [ES]) \, [S]}{[ES]}$$

from which

$$[ES] \, K_m = ([E_0] - [ES]) \, [S]$$

$$[ES] \, K_m = [E_0] \, [S] - [ES] \, [S]$$

$$[ES] \, [S] + [ES] \, K_m = [E_0] \, [S]$$

$$[ES] \, ([S] + K_m) = [E_0] \, [S]$$

$$[ES] = \frac{[E_0] \, [S]}{[S] + K_m}$$

Again, since

$$v_0 = k_2 \, [ES]$$

$$v_0 = \frac{k_2 \, [E_0] \, [S]}{[S] + K_m}$$

and since

$$V_{max} = k_2 \, [E_0]$$

$$v_0 = \frac{V_{max} \, [S]}{[S] + K_m}$$

Hence, the substrate concentration is much greater than the enzyme concentration,

$$[S] \approx [S_0], \text{ so}$$

$$v_0 = \frac{V_{max} \, [S_0]}{[S_0] + K_m} \quad \text{at constant } [E_0]$$

K_m is called the Michaelis constant.

$$\text{When } v_0 = 1/2 \ V_{max}$$

$$\text{then} \quad \frac{V_{max}}{2} = \frac{V_{max} \ [S_0]}{[S_0] + K_m}$$

$$V_{max} \ ([S_0] + K_m) = 2 \ V_{max} \ [S_0]$$

$$[S_0] + K_m = 2 \ [S_0]$$

$$K_m = 2 \ [S_0] - [S_0]$$

$$K_m = [S_0]$$

Therefore, K_m is the value of $[S_0]$ which gives an initial velocity equal to $1/2 \ V_{max}$. Also K_m will have the same units as $[S_0]$.
Thus,

$$v_0 = \frac{a \ [S_0]}{[S_0] + b}$$

where a = maximum value of v_0 (called V_{max})
b = value of $[S_0]$ where $v_0 = 1/2 \ V_{max}$.

Significance of Michaelis-Menten equation

For reaction of the type:

$$E + S \underset{k_{-1}}{\overset{k_1}{\rightleftharpoons}} ES \overset{k_2}{\longrightarrow} EP \overset{k_3}{\longrightarrow} P + E$$

$k_2 = k_{cat}$ where constant k_{cat}, called the turnover number is defined as the maximum no. of substrate molecules which can be converted to products per molecule of enzyme per unit time. The turnover number for most enzymes lies in the range 1-104/second.

We know that:

$$v_0 = \frac{[E_0] \ [S]}{[S] + K_m}$$

Substituting for E_0 using $[E_0] = [E] + [ES]$ and then for $[ES]$ using $K_m = [E] \ [S] / [ES]$

$$v_0 = \frac{k_2 \ [E] \ [S]}{K_m}$$

$$v_0 = \frac{k_{cat} [E] \ [S]}{K_m}$$

The term k_{cat} / K_m is the catalytic efficiency.

The Michaelis-Menten equation has been found to be applicable in determining the constants V_{max} and K_m for many enzyme-catalysed reactions. V_{max} varies with the total concentration of enzyme present but K_m is independent of enzyme concentration and is characteristic of the system being investigated.

In most cases, K_m values lie in the range used to identify a particular enzyme. Generally, K_m values lie in the range $10^{-2} - 10^{-6}$ mol/L. Also, k_1 and k_{-1} are much larger than k_2, where this is the case, the Michaelis-Menten equilibrium is valid and $K_m \cong K_s$.

In general, K_s gives an indication of the affinity of the enzyme for the substrate. A low K_s value indicates a high affinity of enzyme for substrate whereas, a high K_s value indicates a low affinity.

From a practical viewpoint, a knowledge of the K_m of an enzyme is valuable when assaying that enzyme. Enzyme assays should be performed with the enzyme fully saturated with substrate. From the Michaelis-Menten equation, saturation is approached tangentially and only actually achieved at a substrate concentration of infinity. Nevertheless, if the K_m is known, the equation enables an initial substrate concentration to be calculated which, for all practical purposes, can be regarded as saturating the enzyme. For example, if $[S_0] = 100 \ K_m$ then $v_0 = 0.99 \ V_{max}$ irrespective of the actual enzyme concentration provided it is much less than the substrate concentration.

$$v_0 = \frac{V_{max} \ [S_0]}{[S_0] + K_m}$$

$$v_0 = \frac{V_{max} \ (100 \ K_m)}{100 \ K_m + K_m}$$

$$v_0 = \frac{V_{max} \times (100 \ K_m)}{101 \ K_m}$$

$$v_0 = 0.99 \ V_{max}$$

The Lineweaver-Burk plot

The graph of the Michaelis-Menten equation, v_0 against $[S_0]$ is not entirely satisfactory for the determination of V_{max} and K_m. Unless, after a series of experiments, there are at least three consistent points on the plateau of the curve at different $[S_0]$ values, then an accurate value of V_{max}, and hence of K_m, cannot be obtained: the graph, being a curve, cannot be accurately extrapolated upwards from non-saturating values of $[S_0]$.

Lineweaver and Burk overcame this problem without making any fresh assumptions. They simply took the Michaelis-Menten equation:

$$v_0 = \frac{V_{max} [S_0]}{[S_0] + K_m}$$

and inverted it:

$$\frac{1}{v_0} = \frac{[S_0] + K_m}{V_{max}[S_0]} = \frac{[S_0]}{V_{max} [S_0]} + \frac{K_m}{V_{max} [S_0]}$$

Therefore,

$$\frac{1}{v_0} = \frac{K_m}{V_{max}} \times \frac{1}{[S_0]} + \frac{1}{V_{max}}$$

(The Lineweaver-Burk equation)

This is of the form $y = mx + c$, which is the equation of a straight line graph; a plot of y against x has a slope m and intercept c on the y-axis.

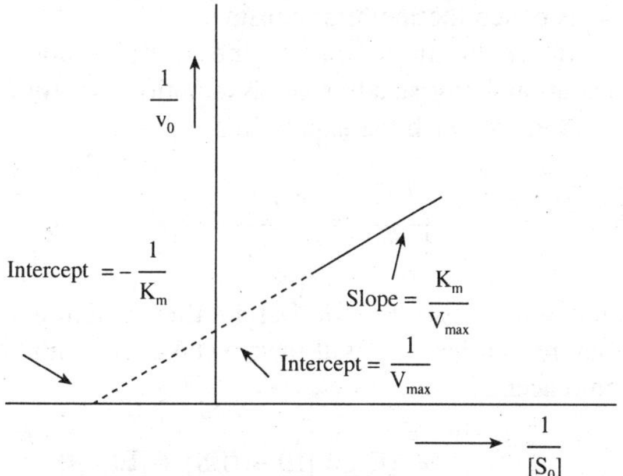

Fig. 6.3 *The Lineweaver-Burk plot.*

6.1.3 Phase-III: The Main Reaction

Figure 6.1 shows that as the reaction proceeds the initial reaction velocity decreases. Tangents drawn to the progress curve at various intervals along the reaction time abscissa will give the instantaneous reaction velocity which will decrease as the reaction proceeds. This occurs even if complete enzyme activity is maintained, and is due to the reaction approaching equilibrium point and the reverse reaction beginning to operate.

6.2   ENZYME INHIBITION

Inhibitors are substances which tend to decrease the rate of an enzyme-catalysed reaction. Although some act on a substrate or cofactor, the discussion here will be restricted to those which combine directly with an enzyme. Reversible inhibitors bind to an enzyme in a reversible fashion and can be removed by dialysis (or simply dilution) to restore full enzyme activity, whereas irreversible inhibitors cannot be removed from an enzyme by dialysis. Sometimes, it may be possible to remove an irreversible inhibitor from an enzyme by introducing another component to the reaction mixture but this would not affect the classification of the original interaction.

Reversible inhibitors, usually rapidly form an equilibrium system with an enzyme to show a definite degree of inhibition (depending on the concentration of enzyme, inhibitor and substrate) which remains constant over the period when initial velocity studies are normally carried out. In contrast, the degree of inhibition by irreversible inhibitors may increase over this period of time.

6.2.1 Reversible Inhibition

6.2.1.1   Competitive inhibition

Competitive inhibitors often closely resemble in some respects, the substrates whose reaction they inhibit and because of this structural similarity they may compete for the same binding-site on the enzyme. The enzyme-bound inhibitor then, either lacks the appropriate reactive group or it is held in an unsuitable position with respect to the catalytic-site of the enzyme or to other potential substrates for a reaction to take place. In either case, a dead-end complex is formed and the inhibitor must dissociate from the enzyme and be replaced by a molecule of substrate before a reaction can take place at that particular enzyme molecule.

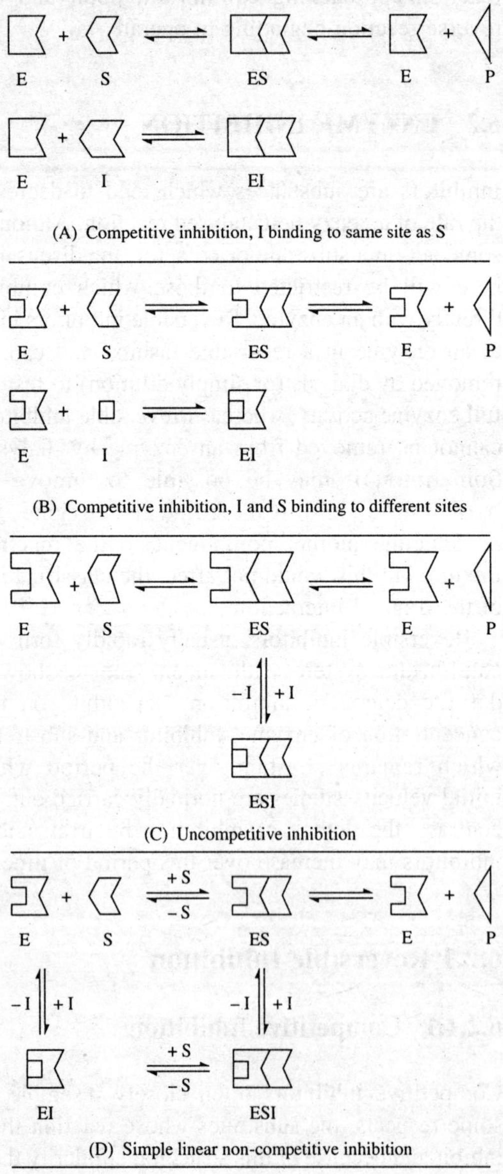

(A) Competitive inhibition, I binding to same site as S

(B) Competitive inhibition, I and S binding to different sites

(C) Uncompetitive inhibition

(D) Simple linear non-competitive inhibition

Fig. 6.4 *Representation of some possible examples of reversible inhibition.*

The effect of a competitive inhibitor depends on the inhibitor concentration, the substrate concentration and the relative affinities of the substrate and the inhibitor for the enzyme. In general, at a particular inhibitor and enzyme concentration, if the substrate concentration is low, the inhibitor will compete favourably with the substrate for the binding sites on the enzyme and the degree of inhibition will be great. However, if at the same inhibitor and enzyme concentration, the substrate concentration is high, then the inhibitor will be much less successful in competing with the substrate for the available binding sites and the degree of inhibition will be less marked.

Let us investigate the steady-state kinetics of a simple single substrate single-binding-site single-intermediate enzyme-catalysed reaction in the presence of a competitive inhibitor, I.

$$E + S \underset{k_{-1}}{\overset{k_1}{\rightleftharpoons}} ES \xrightarrow{k_2} P + E$$

$$-I \updownarrow +I$$

$$EI$$

The dissociation constant for the reaction between E and I is K_i, where

$$K_i = \frac{[E]\,[I]}{[EI]}$$

K_i is called the inhibitor constant.

If we begin to derive the initial velocity equation using steady-state assumption exactly as before, we reach the expression:

$$\frac{[E]\,[S]}{[ES]} = \frac{k_{-1} + k_2}{k_1} = K_m$$

We will substitute for [E] in this equation as before, but here we will have to take the inhibitor into account, since:

$$[E_0] = [E] + [ES] + [EI]$$

$$[E_0] = [E] + [ES] + \frac{[E]\,[I]}{K_i}$$

$$[E_0] = [E] + \frac{[E]\,[I]}{K_i} + [ES]$$

$$[E_0] = [E]\left(1 + \frac{[I]}{K_i}\right) + [ES]$$

$$[E]\left(1 + \frac{[I]}{K_i}\right) = [E_0] - [ES]$$

$$[E] = \frac{[E_0] - [ES]}{\left(1 + \frac{[I]}{K_i}\right)}$$

If we now substitute for [E]:

$$K_m = \frac{\dfrac{[E_0] - [ES]}{\left(1 + \frac{[I]}{K_i}\right)}\,[S]}{[ES]}$$

$$K_m\left(1 + \frac{[I]}{K_i}\right)[ES] = ([E_0] - [ES])\,[S]$$

$$K_m\left(1 + \frac{[I]}{K_i}\right)[ES] = [E_0]\,[S] - [ES]\,[S]$$

$$K_m\left(1 + \frac{[I]}{K_i}\right)[ES] + [ES]\,[S] = [E_0]\,[S]$$

$$[ES]\left[K_m\left(1 + \frac{[I]}{K_i}\right) + [S]\right] = [E_0]\,[S]$$

$$[ES] = \frac{[E_0]\,[S]}{K_m\left(1 + \frac{[I]}{K_i}\right) + [S]}$$

Since,

$$v_0 = k_2\,[ES]$$

$$v_0 = \frac{k_2\,[E_0]\,[S]}{K_m\left(1 + \frac{[I]}{K_i}\right) + [S]}$$

$$v_0 = \frac{V_{max}\,[S]}{[S] + K_m\left(1 + \frac{[I]}{K_i}\right)}$$

Since $[I] \cong [I_0]$ just as $[S] \cong [S_0]$, therefore:

$$v_0 = \frac{V_{max}\,[S_0]}{[S_0] + K_m\left(1 + \frac{[I_0]}{K_i}\right)}$$

This is an equation of the same form as the Michaelis-Menten, the only difference is that K_m has increased by a factor $(1 + [I_0]/K_i)$.

For simple competitive inhibition V_{max} is altered so that,

$$K'_m = K_m\left(1 + \frac{[I_0]}{K_i}\right)$$

Where K'_m is apparent K_m in the presence of an initial concentration $[I_0]$ of competitive inhibitor.

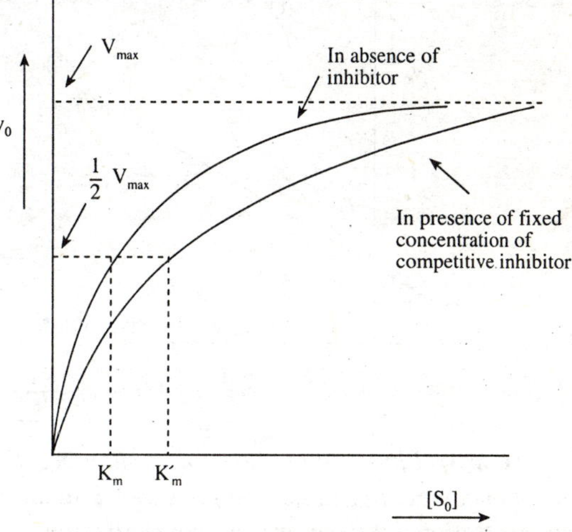

Fig. 6.5 *Michaelis-Menten plot (at fixed $[E_0]$) showing the effect of a competitive inhibitor.*

The Lineweaver-Burk equation of a competitive inhibitor will be :

$$\frac{1}{v_0} = \frac{K'_m}{V_{max}} \times \frac{1}{[S_0]} + \frac{1}{V_{max}}$$

and the Lineweaver-Burk plots showing the effect of competitive inhibition are shown in figure 6.6.

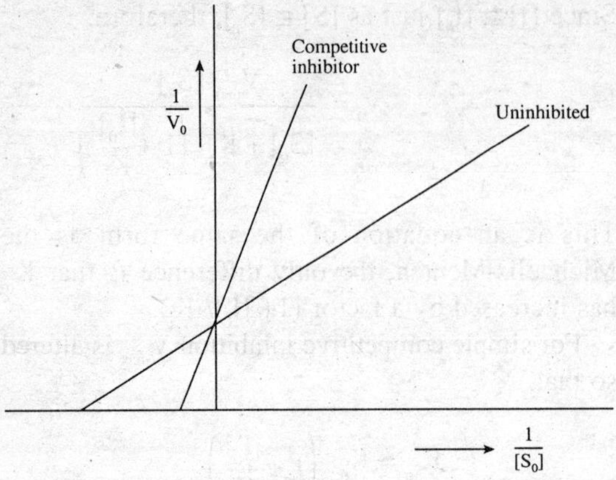

Fig. 6.6(a) *Lineweaver-Burk plot showing the effect of competitive inhibition.*

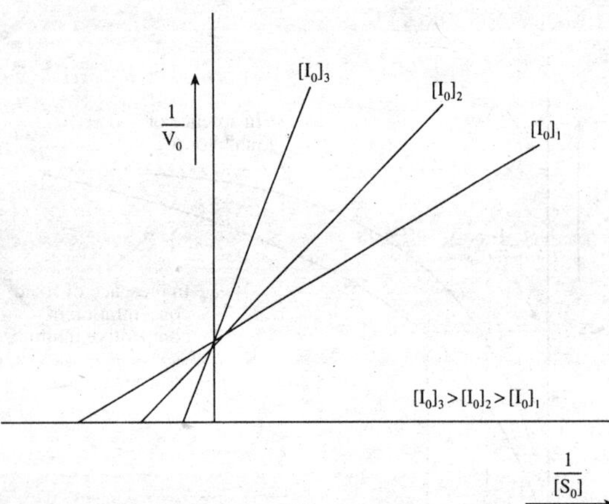

Fig. 6.6(b) *Lineweaver-Burk plot showing the effect of competitive inhibition for several inhibitor concentrations at fixed enzyme concentrations.*

6.2.1.2 Uncompetitive inhibition

Uncompetitive inhibitors bind only to the enzyme-substrate complex and not to the free enzyme. Substrate-binding could cause a conformational change to take place in the enzyme and reveal an inhibitor binding-site or the inhibitor could bind directly to the enzyme-bound substrate. In neither case, the inhibitor competes with the substrate for the same binding site, so the inhibition cannot be overcome by increasing the substrate concentration. Both K_m and V_{max} are altered.

Once again let us consider the simplest situation:

$$E + S \underset{k_{-1}}{\overset{k_1}{\rightleftharpoons}} ES \xrightarrow{k_2} P + E$$

ESI is a dead-end complex; the inhibitor constant $K_i = ([ES]\,[I]\,/\,[ESI])$.

Under steady-state conditions,

$$K_m = \frac{[E]\,[S]}{[ES]}$$

For this system,

$$[E_0] = [E] + [ES] + [ESI]$$

$$[E_0] = [E] + [ES] + \frac{[ES]\,[I]}{K_i}$$

$$[E_0] = [E] + [ES]\left(1 + \frac{[I]}{K_i}\right)$$

$$[E] = [E_0] - [ES]\left(1 + \frac{[I]}{K_i}\right)$$

Substitute for $[E]$ in $K_m = [E]\,[S]\,/[ES]$

$$K_m = \frac{\left[[E_0] - [ES]\left(1 + \frac{[I]}{K_i}\right)\right][S]}{[ES]}$$

$$K_m = \frac{[E_0][S] - [ES][S]\left(1 + \frac{[I]}{K_i}\right)}{[ES]}$$

$$K_m[ES] = [E_0][S] - [ES][S]\left(1 + \frac{[I]}{K_i}\right)$$

$$[E_0][S] = [ES][S]\left(1 + \frac{[I]}{K_i}\right) + K_m[ES]$$

$$[E_0][S] = [ES]\left[[S]\left(1+\frac{[I]}{K_i}\right) + K_m\right]$$

$$[ES] = \frac{[E_0][S]}{\left[[S]\left(1+\frac{[I]}{K_i}\right) + K_m\right]}$$

Substitute for [ES] in $v_0 = k_2\,[ES]$ as before,

$$v_0 = \frac{k_2\,[E_0][S]}{\left[[S]\left(1+\frac{[I]}{K_i}\right) + K_m\right]}$$

$$v_0 = \frac{V_{max}\,[S]}{\left[[S]\left(1+\frac{[I]}{K_i}\right) + K_m\right]}$$

Since $[I] \cong [I_0]$ just as $[S] \cong [S_0]$, therefore:

$$v_0 = \frac{V_{max}\,[S_0]}{\left[[S_0]\left(1+\frac{[I_0]}{K_i}\right) + K_m\right]}$$

Dividing throughout by $(1 + [I_0] / K_i)$ gives:

$$v_0 = \frac{\dfrac{V_{max}}{\left(1+\dfrac{[I_0]}{K_i}\right)}[S_0]}{[S_0] + \dfrac{K_m}{\left(1+\dfrac{[I_0]}{K_i}\right)}}$$

This is an equation of the same form as of the Michaelis-Menten, the constants K_m and V_{max} both being divided by a factor $[1 + ([I_0] / K_i)]$.

Thus for uncompetitive inhibition,

$$V'_{max} = \frac{V_{max}}{\left(1+\dfrac{[I_0]}{K_i}\right)} \quad \text{and} \quad K'_m = \frac{K_m}{\left(1+\dfrac{[I_0]}{K_i}\right)}$$

Where V'_{max} is the value of V_{max} in the presence of an initial concentration $[I_0]$ of uncompetitive inhibitor and K'_m is the apparent value of K_m under the same conditions. An inhibitor concentration equal to K_i will halve the values of both V_{max} and K_m.

The Lineweaver equation, in the presence of an uncompetitive inhibitor is:

$$\frac{1}{v_0} = \frac{K'_m}{V'_{max}} \times \frac{1}{[S_0]} + \frac{1}{V_{max}}$$

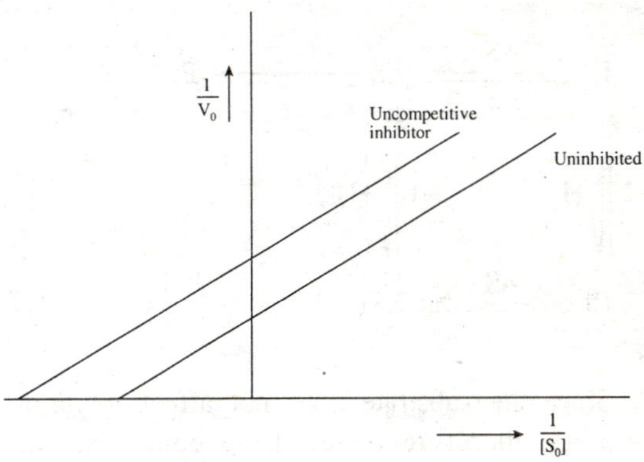

Fig. 6.7(a) *Lineweaver-Burk plots showing the effect of uncompetitive inhibition.*

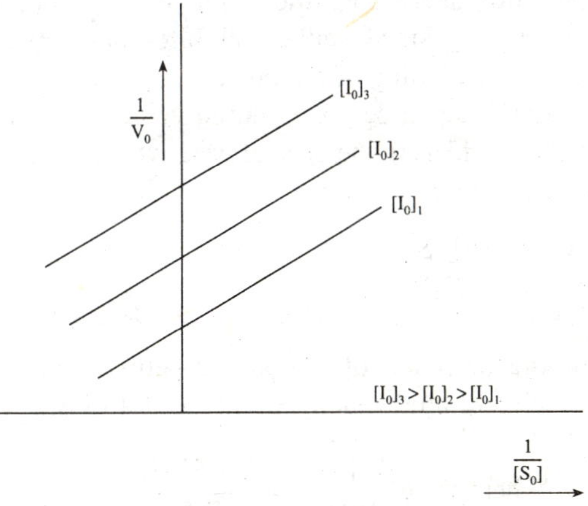

Fig. 6.7(b) *Lineweaver-Burk plots showing the effect of uncompetitive inhibition for several inhibitor concentrations at fixed enzyme concentration.*

6.2.1.3 Non-competitive inhibition

A non-competitive inhibitor can combine with an enzyme molecule to produce a dead-end complex, regardless of whether a substrate molecule is bound or not. Hence, the inhibitor must bind at a different site from the substrate. We shall only consider the case, where the inhibitor destroys the catalytic activity of the enzyme, either by binding to the catalytic site or as a result of a conformational change affecting the catalytic site but does not affect substrate-binding. The situation for a simple single-substrate reaction will be as follows:

$$E \; \underset{-S}{\overset{+S}{\rightleftharpoons}} \; ES \longrightarrow P$$

$$EI \; \underset{-S}{\overset{+S}{\rightleftharpoons}} \; ESI$$

Since the substrate does not affect inhibitor-binding, therefore under these conditions the reactions $E + I \rightleftharpoons EI$ and $ES + I \rightleftharpoons ESI$ have an identical dissociation constant K_i, again called the inhibitor constant. The total enzyme concentration is effectively reduced by the inhibitor decreasing the value of V_{max} but not altering K_m since neither inhibitor nor substrate affect the binding of the other.

Let us again derive an initial velocity equation, only considering the special case where $K_m \cong K_s$.

As before,

$$K_m = \frac{[E] \, [S]}{[ES]}$$

In the presence of a non-competitive inhibitor which will bind equally well to E or to ES,

i.e. where $K_i = \dfrac{[E] \, [I]}{[EI]} = \dfrac{[ES] \, [I]}{[ESI]}$

For this system,

$$[E_0] = [E] + [ES] + [EI] + [ESI]$$

$$[E_0] = [E] + [ES] + \frac{[E] \, [I]}{K_i} + \frac{[ES] \, [I]}{K_i}$$

$$[E_0] = [E] + \frac{[E] \, [I]}{K_i} + [ES] + \frac{[ES] \, [I]}{K_i}$$

$$[E_0] = [E] \left(1 + \frac{[I]}{K_i}\right) + [ES] \left(1 + \frac{[I]}{K_i}\right)$$

$$[E_0] = ([E] + [ES]) \left(1 + \frac{[I]}{K_i}\right)$$

$$[E] + [ES] = \frac{[E_0]}{\left(1 + \dfrac{[I]}{K_i}\right)}$$

$$[E] = \frac{[E_0]}{\left(1 + \dfrac{[I]}{K_i}\right)} - [ES]$$

Substituting [E] in $K_m = \dfrac{[E] \, [S]}{[ES]}$

$$K_m = \frac{\left[\dfrac{[E_0]}{\left(1 + \dfrac{[I]}{K_i}\right)} - [ES]\right][S]}{[ES]}$$

$$K_m \, [ES] = \frac{[E_0]}{\left(1 + \dfrac{[I]}{K_i}\right)} [S] - [ES] \, [S]$$

$$K_m \, [ES] + [ES] \, [S] = \frac{[E_0]}{\left(1 + \dfrac{[I]}{K_i}\right)} [S]$$

$$[ES] = \frac{[E_0]}{\left(1 + \dfrac{[I]}{K_i}\right)} \times \frac{[S]}{(K_m + [S])}$$

Since $v_0 = k_2 \, [ES]$, therefore

$$v_0 = k_2 \frac{[E_0]}{\left(1 + \dfrac{[I]}{K_i}\right)} \times \frac{[S]}{(K_m + [S])}$$

Taking $[I] = [I_0]$ and $[S] = [S_0]$

$$v_0 = k_2 \frac{[E_0]}{\left(1 + \dfrac{[I_0]}{K_i}\right)} \times \frac{[S_0]}{(K_m + [S_0])}$$

or

$$v_0 = \frac{V_{max}}{\left(1 + \dfrac{[I_0]}{K_i}\right)} \times \frac{[S_0]}{(K_m + [S_0])}$$

$$(\because k_2 [E_0] = V_{max})$$

This is the form of the Michaelis-Menten equation, with V_{max} being divided by a factor $[1 + ([I_0]/K_i)]$.

Thus, for simple linear non-competitive inhibition, K_m is unchanged and V_{max} is altered so that,

$$V_{max}' = \frac{V_{max}}{\left(1 + \dfrac{[I_0]}{K_i}\right)}$$

The Lineweaver equation, for simple linear non-competitive inhibition is:

$$\frac{1}{v_0} = \frac{K_m}{V_{max}'} \times \frac{1}{[S_0]} + \frac{1}{V_{max}'}$$

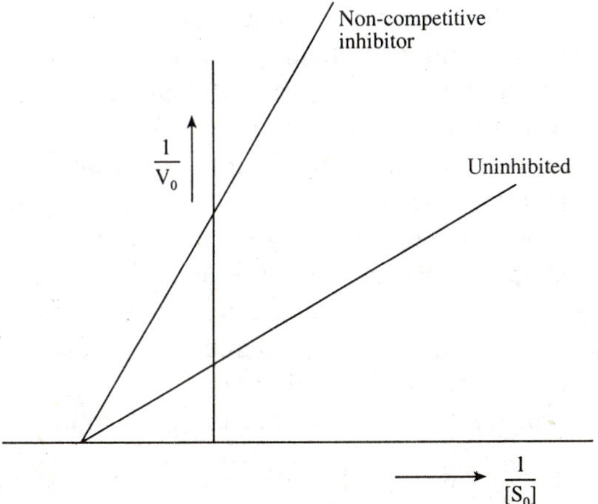

Fig. 6.8(a) *Lineweaver-Burk plots showing the effect of non-competitive inhibition.*

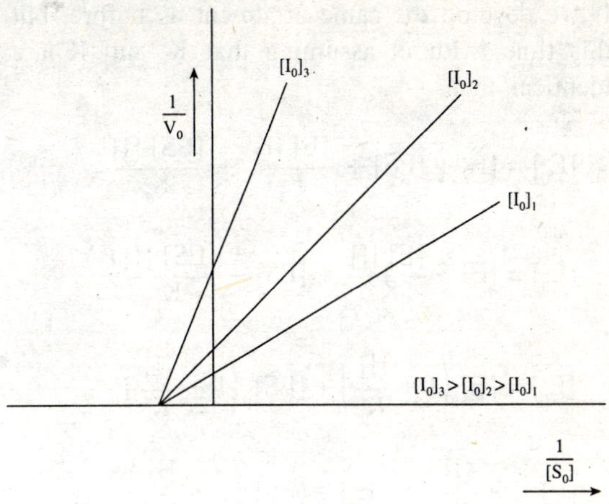

Fig. 6.8(b) *Lineweaver-Burk plots showing the effect of non-competitive inhibition for several inhibitor concentrations at fixed enzyme concentration.*

6.2.1.4 Mixed inhibition

In the previous section, we obtained an expression for simple linear non-competitive inhibition which depended on the equilibrium-assumption being valid and further assumed that substrate-binding and inhibitor-binding were completely independent. Let us now consider the situation where the second assumption is not made.

There are two processes by which the inhibitor may bind to the enzyme:

$$E + I \rightleftharpoons EI \text{ (inhibitor constant } K_i)$$
$$\text{and } ES + I \rightleftharpoons ESI \text{ (inhibitor constant } K_I)$$

Hence,

$$K_i = \frac{[E][I]}{[EI]} \text{ and } K_I = \frac{[ES][I]}{[ESI]}$$

As before, for a single-substrate reaction:

$$K_m = \frac{[E][S]}{[ES]}$$

and

$$[E_0] = [E] + [ES] + [EI] + [ES I]$$

If we develop the same argument as before, but this time, without assuming that K_I and K_i are identical, then,

$$[E_0] = [E] + [ES] + \frac{[E]\,[I]}{K_i} + \frac{[ES]\,[I]}{K_I}$$

$$[E_0] = [E] + \frac{[E]\,[I]}{K_i} + [ES] + \frac{[ES]\,[I]}{K_I}$$

$$[E_0] = [E]\left(1 + \frac{[I]}{K_i}\right) + [ES]\left(1 + \frac{[I]}{K_I}\right)$$

$$[E]\left(1 + \frac{[I]}{K_i}\right) = [E_0] - [ES]\left(1 + \frac{[I]}{K_I}\right)$$

$$[E] = \frac{[E_0] - [ES]\left(1 + \dfrac{[I]}{K_I}\right)}{\left(1 + \dfrac{[I]}{K_i}\right)}$$

Substituting for $[E]$ in the expression for K_m:

$$K_m = \frac{\left[\dfrac{[E_0] - [ES]\left(1 + \dfrac{[I]}{K_I}\right)}{\left(1 + \dfrac{[I]}{K_i}\right)}\right][S]}{[ES]}$$

$$K_m[ES]\left(1 + \frac{[I]}{K_i}\right) = [E_0]\,[S] - [ES]\,[S]\left(1 + \frac{[I]}{K_I}\right)$$

$$[E_0]\,[S] = [ES]\,[S]\left(1 + \frac{[I]}{K_I}\right) + K_m\,[ES]\left(1 + \frac{[I]}{K_i}\right)$$

$$[E_0]\,[S] = [ES]\left[[S]\left(1 + \frac{[I]}{K_I}\right) + K_m\left(1 + \frac{[I]}{K_i}\right)\right]$$

$$[ES] = \frac{[E_0]\,[S]}{[S]\left(1 + \dfrac{[I]}{K_I}\right) + K_m\left(1 + \dfrac{[I]}{K_i}\right)}$$

Continuation as before gives:

$$v_0 = \frac{k_2\,[E_0]\,[S]}{[S]\left(1 + \dfrac{[I]}{K_I}\right) + K_m\left(1 + \dfrac{[I]}{K_i}\right)}$$

$$v_0 = \frac{V_{max}\,[S_0]}{[S_0]\left(1 + \dfrac{[I_0]}{K_I}\right) + K_m\left(1 + \dfrac{[I_0]}{K_i}\right)}$$

If numerator and denominator are both divided by $[1 + ([I_0] / K_I)]$,

$$v_0 = \frac{\dfrac{V_{max}}{\left(1 + \dfrac{[I_0]}{K_I}\right)}[S_0]}{[S_0] + \dfrac{K_m\left(1 + \dfrac{[I_0]}{K_i}\right)}{\left(1 + \dfrac{[I_0]}{K_I}\right)}}$$

This is the same form as the Michaelis-Menten equation and can be written as:

$$v_0 = \frac{V'_{max}\,[S_0]}{[S_0] + K'_m}$$

Where,

$$V'_{max} = \frac{V_{max}}{\left(1 + \dfrac{[I_0]}{K_I}\right)} \quad \text{and} \quad K'_m = K_m\frac{\left(1 + \dfrac{[I_0]}{K_i}\right)}{\left(1 + \dfrac{[I_0]}{K_I}\right)}$$

The equation derived above is a relatively general one, since no assumptions were made about the values of K_i and K_I, and it can be simplified for special cases. If ESI cannot be formed, then $K_I = \infty$ and the equation becomes that for competitive inhibition, regardless of whether the substrate and inhibitor bind to the same or different sites. If the complex ESI can occur but not EI, then $K_i = \infty$ and the equation simplifies to that for uncompetitive inhibition. When $K_i = K_I$, the equation reduces to that for linear non-competitive inhibition.

Similarly, the Lineweaver equation inhibition is:

$$\frac{1}{v_0} = \frac{K'm}{V'_{max}} \times \frac{1}{[S_0]} + \frac{1}{V'_{max}}$$

In the situations where $K_I > K_i$, the plots cross to the left of the $1/v_0$ axis but above the $1/[S_0]$ axis. This situation has been termed competitive-noncompetitive inhibition, because the pattern observed lies between those for competitive and non-competitive inhibition.

In the situations where $K_I < K_i$, the plots cross to the left of the $1/v_0$ axis and below the $1/[S_0]$ axis. This form of mixed inhibition has been termed non-competitive-uncompetitive inhibition, because the pattern is intermediate between those for non-competitive and uncompetitive inhibition.

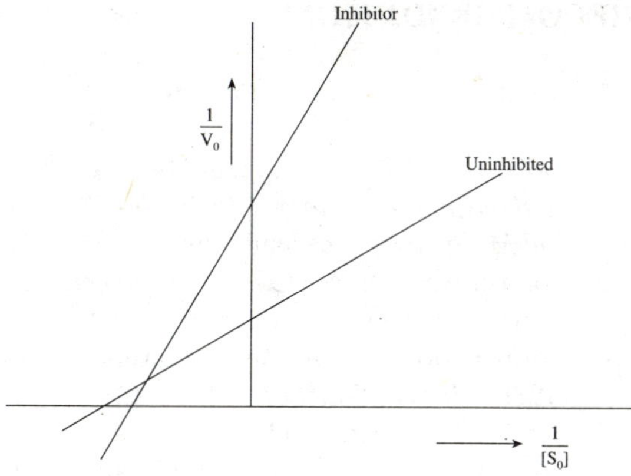

Fig. 6.9(a) *Lineweaver-Burk plots showing the effect of mixed inhibition when $K_I > K_i$.*

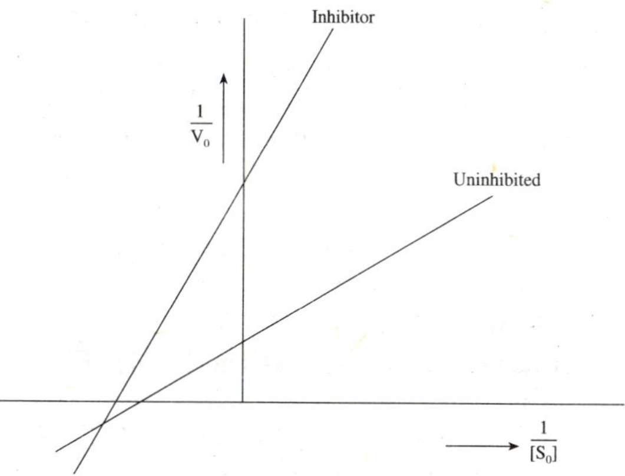

Fig. 6.9(b) *Lineweaver-Burk plots showing the effect of mixed inhibition when $K_I < K_i$.*

6.2.2 Irreversible Inhibition

An irreversible inhibitor binds to the active site of the enzyme by an irreversible reaction:

$$E + I \longrightarrow EI$$

and hence, cannot subsequently dissociate from it. A covalent bond is usually formed between inhibitor and enzyme. The inhibitor may act by preventing substrate-binding or it may destroy some component of the catalytic site. Compounds which irreversibly denature the enzyme protein or cause non-specific inactivation of the active site are not usually regarded as irreversible inhibitors.

Unlike reversible inhibition, where an equilibrium is quickly set up between inhibitor and enzyme, making the system suitable for investigation by initial velocity studies, irreversible inhibition is progressive and will increase with time until either all the inhibitor or all the enzyme present has been used up in forming enzyme-inhibitor complex.

Regardless of whether the reaction between enzyme and irreversible inhibitor has gone to completion before initial velocity studies are commenced, these will give little or no information about the characteristics of the inhibitor. It is much more useful to describe the system in terms of the rate constant for the binding of inhibitor to substrate. If these react on a 1:1 basis then the time taken to reduce the enzyme activity to 50 per cent of its original value will be inversely proportional to the initial inhibitor concentration (at fixed total enzyme concentration).

Irreversible inhibitors effectively reduce the concentration of enzyme present. An inhibitor of initial concentration $[I_0]$ will reduce the concentration of active enzyme from an initial value of $[E_0]$ to $[E_0] - [I_0]$, assuming the inhibitor is not in excess. If a substrate is introduced after the reaction between inhibitor and enzyme has gone to completion, a system which obeys the Michaelis-Menten equation in the absence of inhibitor will still do so. The value of K_m will be the same as for the uninhibited reaction, but V_{max} will be reduced to V'_{max}.

In the absence of inhibitor, $V_{max} = k_{cat} [E_0]$.
In the presence of inhibitor, $V'_{max} = k_{cat} ([E_0]-[I_0])$

$$\frac{V'_{max}}{V_{max}} = \frac{[E_0] - [I_0]}{[E_0]}$$

$$V'_{max} = V_{max} [E_0] \left(1 - \frac{[I_0]}{[E_0]} \right)$$

Therefore, patterns resembling those for reversible non-competitive inhibition, with unchanged K_m and reduced V_{max}, may be obtained with irreversible inhibitors, even when the inhibitor binds to the same site as the substrate. Hence, if a pattern of non-competitive inhibition is obtained in the investigation of a system, it is important to establish whether the inhibition is reversible or irreversible before the results can be interpreted.

SUMMARY

The hyperbolic graphs of v_0 against $[S_0]$ obtained experimentally for many enzyme-catalysed reactions can be obtained by the Michaelis-Menten equation. The original equilibrium-assumption of Michaelis and Menten is now seen as a special case of the steady-state assumption of Briggs and Haldane. Steady-state kinetics can be used to calculate K_m and k_{cat}, which are characteristic of a particular enzyme. These calculations are facilitated by the use of linear plots derived from the Michaelis-Menten equation.

The kinetic mechanism of a reaction can be investigated in more detail by the use of rapid-reaction techniques.

Competitive inhibitors usually compete with the substrate for the same binding site on the enzyme. In the characteristic form, Michaelis-Menten kinetics are obeyed, K_m is increased and V_{max} unchanged. Uncompetitive inhibitors bind to a site other than the substrate-binding site on the enzyme-substrate complex, altering the K_m and V_{max} but not the slope of the Lineweaver-Burk plot. Non-competitive inhibitors bind to a site other than the substrate-binding site on the enzyme and enzyme-substrate complex. In the characteristic form, Michaelis-Menten kinetics are obeyed, K_m is unchanged and V_{max} decreased.

Forms of inhibition obeying Michaelis-Menten kinetics but not giving patterns characteristic of competitive, uncompetitive or non-competitive inhibition are usually termed mixed inhibition, irrespective of the actual mechanism. All these forms of inhibition are reversible, but irreversible inhibition is also known. Irreversible inhibitors have been used to identify amino acids in the active centres of enzymes.

RECOMMENDATIONS

1. Brodbeck, U. (1980), *Enzyme Inhibitors*, Verlag-Chemie.
2. Colowick, S.P. and Kaplan, N.O. (1979), *Methods in Enzymology, 63: Enzyme Kinetics and Mechanism*, Academic Press.
3. Cornish-Bowden, A. (1979), *Fundamentals of Enzyme Kinetics*, Butterworth.
4. Cornish-Bowden, A. and Wharton, C.W. (1988), *Enzyme Kinetics*, IRL Press.
5. Dawes, E.A. (1980), *Quantitative Problems in Biochemistry*, 6th edn. (Chapter 6), Longmam.
6. Dixon, M., Webb, E. C., Thorne, C.J.R and Tipton, K.F. (1979), *Enzymes*, 3rd edn. (Chapters 4 and 8), Longman.
7. Halford, S.E. (1974), Rapid Reaction Techniques, in Bull., A.T., Lagnado, J.R., Thomas, J.D. and Tipton, C.J.R. and Tipton, K.F. *Companion to Biochemistry*, Longman.
8. Hammes, G. G. (1982), *Enzyme Catalysis and Regulation*, Academic Press.
9. Jain, M.K. (1982), *Handbook of Enzyme Inhibitors*, John Wiley.
10. Morris, J.G. (1974), *A Biologist's Physical Chemistry*, 2nd edn. (Chapter 11), Edward Arnold.
11. Price, N.C. and Stevens, L. (1989), *Fundamentals of Enzymology*, 2nd edn. (Chapter 4), Oxford University Press.
12. Zollner, H. (1990), *Handbook of Enzyme Inhibitors*, VCH.

7 Kinetics of Multi-Substrate Enzyme Catalysed Reactions

7.1 ORDER OF SUBSTRATE BINDING

In the reaction of an enzyme with two substrates, the binding of the substrates can occur sequentially in a specific order or it may be non-sequential.

7.1.1 Sequential Mechanisms

Mechanisms where all substrates must add to the enzyme before any products are released are called sequential. Thus the two-substrate reactions which follow sequential mechanisms essentially proceed through a ternary adsorption complex (enzyme + two substrates).

A ternary enzyme-substrate complex can be formed in two ways. The substrate, are bound to the enzyme in a random fashion (random mechanism) or they are bound in a well-defined order (ordered mechanism).

7.1.1.1 Random mechanism

A random mechanism is one in which any substrate can bind first to the enzyme and any product can leave first. It is a sequential mechanism and for a two-substrate reaction involves the formation of a ternary complex (involving one enzyme and both substrates).

$$
\begin{array}{ccc}
E + AX \rightleftharpoons E.AX & & EA \rightleftharpoons E + A \\
\quad\searrow{+B} & & \nearrow{-BX} \\
& E.AX.B \rightleftharpoons E.A.BX & \\
\quad\nearrow{+AX} & & \searrow{-A} \\
E + B \rightleftharpoons EB & & E.BX \rightleftharpoons E + BX
\end{array}
$$

There will be two substrate binding sites on the enzyme, one for A/AX and one for B/BX.

7.1.1.2 Ordered mechanism

An ordered-mechanism is a sequential mechanism where the order of binding to and leaving the enzyme is compulsory. For a two-substrate reaction, a ternary complex will be involved. The precise order of mechanism must be specified, e.g.

$$E + AX \rightleftharpoons E.AX \xrightarrow{+B} E.AX.B \rightleftharpoons E.A.BX \xrightarrow{-BX} EA \rightleftharpoons E + A$$

or

$$E + B \rightleftharpoons EB \xrightarrow{+AX} E.B.AX \rightleftharpoons E.BX.A \xrightarrow{-A} E.BX \rightleftharpoons E + BX$$

As before, the enzyme will have a binding site for A/AX and a separate one for B/BX.

7.1.2 Non-sequential Mechanisms

An example of non-sequential mechanism is the ping-pong bi-bi or double displacement mechanism:

$$
\begin{aligned}
E + AX &\rightleftharpoons E.AX \rightleftharpoons EX.A \rightleftharpoons EX + A \\
EX + B &\rightleftharpoons EX.B \rightleftharpoons E.XB \rightleftharpoons E + BX
\end{aligned}
$$

AX first binds to the enzyme E, forming a binary complex E.AX (X is usually a small group and does not participate in the reaction as a free molecule, so it is not regarded as a separate reactant). An intramolecular reorganization takes place, the bond E-X being formed and the X-A bond being broken. The first product 'A' then leaves before the second substrate arrives. B cannot bind to the enzyme E but can bind to the modified enzyme EX. Since only one substrate is present on the enzyme at any one time, there may only be a single binding site. Another intramolecular rearrangement takes place, the bond B-X being formed and the bond E-X being broken. The second product 'BX' is then liberated leaving the enzyme in its original form.

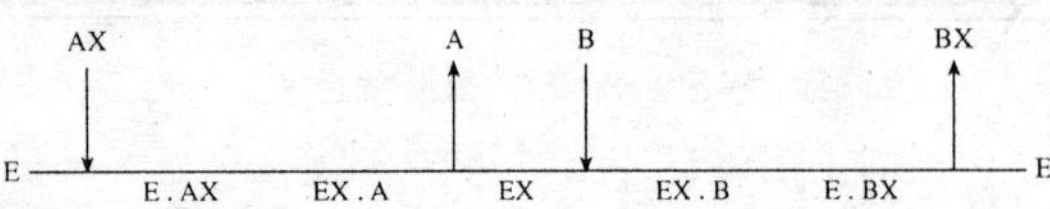

7.2 STEADY-STATE KINETICS

7.2.1 The General Rate Equation of Alberty

Many two-substrate enzyme-catalysed reactions obey the Michaelis-Menten equation with respect to one substrate at constant concentration of the other substrates. This applies both to reactions catalysed by enzymes with just one binding site per substrate and by those with several binding sites per substrate, provided there is no interaction between the binding sites. For such reactions, Alberty derived the general equation:

$$v_0 = \frac{V_{max} [AX_0] [B_0]}{K_m^B [AX_0] + K_m^{AX} [B_0] + [AX_0][B_0] + K_s^{AX} \ K_m^B}$$

Where,

V_{max} is the maximum possible v_0 when AX and B are both saturating.

K_m^{AX} is the concentration of AX which gives $1/2 \ V_{max}$ when B is saturating.

K_m^B is the concentration of B which gives $1/2 \ V_{max}$ when AX is saturating.

K_s^{AX} is the dissociation constant for
$$E + AX \rightleftharpoons EAX$$

The total enzyme concentration is constant and much smaller than the concentration of the two substrates.

At very large $[B_0]$ the general equation simplifies to:

$$v_0 = \frac{V_{max} [AX_0] [B_0]}{K_m^{AX} [B_0] + [AX_0] [B_0]}$$

$$v_0 = \frac{V_{max} [AX_0] [B_0]}{[B_0] (K_m^{AX} + [AX_0])}$$

$$v_0 = \frac{V_{max} [AX_0]}{K_m^{AX} + [AX_0]}$$

$$v_0 = \frac{V_{max}}{1 + \dfrac{K_m^{AX}}{[AX_0]}}$$

$$v_0 = \frac{V_{max}}{\dfrac{[AX_0] + K_m^{AX}}{[AX_0]}}$$

$$v_0 = \frac{V_{max} [AX_0]}{[AX_0] + K_m^{AX}}$$

which is the Michaelis-Menten equation.

Similarly at very large $[AX_0]$

$$v_0 = \frac{V_{max} [AX_0] [B_0]}{K_m^B [AX_0] + [AX_0] [B_0]}$$

$$v_0 = \frac{V_{max} [AX_0] [B_0]}{[AX_0] (K_m^B + [B_0])}$$

$$v_0 = \frac{V_{max} [B_0]}{K_m^B + [B_0]}$$

$$v_0 = \frac{V_{max}}{1 + \dfrac{K_m^B}{[B_0]}}$$

which is the Michaelis-Menten equation.

It should be noted that the general rate equation of Alberty will not be the same for all sorts of multi-substrate mechanisms. So, for a compulsory-order mechanism where B binds first to the enzyme, the $K_s^{AX} K_m^B$ term would be replaced in the general equation by $K_s^B K_m^{AX}$.

$$v_0 = \frac{V_{max}\,[AX_0]\,[B_0]}{K_m^B\,[AX_0] + K_m^{AX}\,[B_0] + [AX_0][B_0] + K_s^B\,K_m^{AX}}$$

For a random-order mechanism either term could be used and for a ping-pong bi-bi mechanism, the liberation of A from the enzyme in the initial period of the reaction will be irreversible because the concentration of product A will be negligible. Hence K_s^{AX} and so $K_s^{AX}.\,K_m^B$ will be zero, giving this mechanism a simple rate equation as:

$$v_0 = \frac{V_{max}\,[AX_0]\,[B_0]}{K_m^B\,[AX_0] + K_m^{AX}\,[B_0] + [AX_0]\,[B_0]}$$

7.2.2 Investigation of Reaction Mechanisms

7.2.2.1 The use of Lineweaver-Burk plot

Reactions which obey the general rate equation of Alberty give linear plots of $1/v_0$ against $1/[AX_0]$ at constant $[B_0]$ and of $1/v_0$ against $1/[B_0]$ at constant $[AX_0]$. The plots for such reactions, e.g. those proceeding by compulsory-order or random-order ternary-complex mechanisms will be of the form shown in Figure 7.1(a).

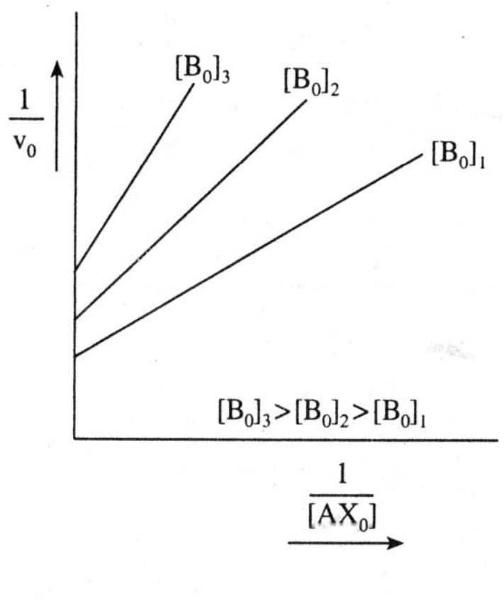

(a)

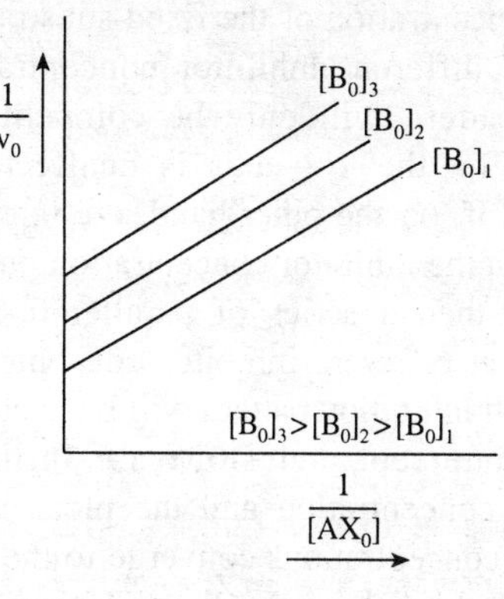

(b)

Fig. 7.1 *Primary plots obtained at non-saturating concentration of both substrates,*
(a) for compulsory-order and random-order ternary-complex mechanisms;
(b) for ping-pong bi-bi mechanisms.
(In each case similar results would be obtained for plots of $1/v_0$ against $1/[B_0]$ at fixed $[AX_0]$.)

In contrast, the equation for reactions proceeding by a ping-pong mechanism is simpler in that $K_s^{AX} K_m^B = 0$. A series of the plots obtained at different concentrations of the fixed substrate would thus be parallel if a ping-pong bi-bi mechanism operates (Figure 7.1b).

Compulsory-order and random-order ternary-complex mechanisms can thus be distinguished from ping-pong bi-bi mechanisms, but not from each other, by the use of these plots.

7.2.2.2 The use of inhibitors which compete with substrates for binding sites

The use of inhibitors which compete with one of the substrates for a site on the enzymes can give useful information on the mechanism of the reaction. A product of the reaction, for example, if present at the start may compete with a substrate for a binding site on the enzyme and thus slow down the rate of the forward reaction.

Because of the complex nature of two-substrate reactions, the fact that a particular inhibitor acts in a competitive way will not necessarily result in a characteristic competitive inhibition pattern. At a single concentration of the fixed substrate but at a series of different inhibitor concentrations, the overall pattern will only be competitive if the intercept on the $1/v_0$ axis is unaffected by the inhibitor. If, on the other hand, the intercept does depend on the inhibitor concentration but the slope does not then a series of parallel lines will be obtained at different inhibitor concentrations and the overall inhibition pattern will be uncompetitive. If both intercept and slope are influenced by inhibitor concentration and the plots at different inhibitor concentrations converge to the left of the $1/v_0$ axis then the overall inhibition pattern is mixed, except in the special case of non-competitive inhibition where the plots converge at the horizontal axis. All of these inhibition patterns may be seen with two-substrate reactions where the inhibition is essentially competitive in nature.

Let us look first at a random-order ternary-complex mechanism,

$$E + AX \rightleftharpoons E.AX \underset{+B}{\rightleftharpoons} E.AX.B \rightleftharpoons E.A.BX$$
$$E + B \rightleftharpoons EB \underset{+AX}{\rightleftharpoons} E.AX.B$$
$$E.A.BX \underset{-BX}{\rightleftharpoons} EA \rightleftharpoons E + A$$
$$E.A.BX \underset{-A}{\rightleftharpoons} E.BX \rightleftharpoons E + BX$$

If the forward reaction is investigated under steady-state conditions at fixed $[AX_0]$ and variable $[B_0]$ in the presence of the product BX, then a competitive inhibition pattern will result, the variable substrate B will compete with BX for the same site on the same enzyme-form (E). Similar results will be obtained for inhibition by the product A at fixed $[B_0]$ and variable $[AX_0]$.

For a compulsory-order mechanism where AX binds first

$$E+AX \rightleftharpoons E.AX \underset{+B}{\rightleftharpoons} E.AX.B \rightleftharpoons E.A.BX \underset{-BX}{\rightleftharpoons} EA \rightleftharpoons E+A$$

A and AX will compete for a site on the enzyme-form E, therefore, overall competitive inhibition will result from inhibition by the product A where AX is the variable substrate. No other combination of inhibitor and variable substrate will give a pattern characteristic of competitive inhibition for this mechanism. Inhibition by the product BX where B is the variable substrate will give mixed inhibition since BX and B do not bind to the same enzyme-form, BX binding to EA and B to E.AX.

Similarly, for a compulsory-order mechanism of the form:

$$E+B \rightleftharpoons EB \underset{+AX}{\rightleftharpoons} E.B.AX \rightleftharpoons E.BX.A \underset{-A}{\rightleftharpoons} E.BX \rightleftharpoons E+BX$$

only inhibition by the product BX where B is the variable substrate will give a pattern characteristic of competitive inhibition.

For a ping-pong bi-bi mechanism,

$$E + AX \rightleftharpoons E.AX \rightleftharpoons EX.A \rightleftharpoons EX + A$$
$$EX + B \rightleftharpoons EX.B \rightleftharpoons E.XB \rightleftharpoons E + BX$$

AX and BX compete for a site on the enzyme-form E, whereas A and B compete for a site on the enzyme-form EX. Therefore inhibition by BX when AX is the variable substrate will give overall competitive inhibition, as will inhibition by A when B is the variable substrate.

The overall inhibition patterns under these conditions may be summarized as follows:

Mechanism	Product used as inhibitor	Inhibition pattern observed	
		With respect to varying $[AX_0]$	With respect to varying $[B_0]$
Compulsory-order ternary-complex (AX binding first)	A	competitive	mixed
	BX	mixed	mixed
Compulsory-order ternary-complex (B binding first)	A	mixed	mixed
	BX	mixed	competitive
Random-order ternary-complex	A	competitive	competitive or mixed
	BX	competitive or mixed	competitive
ping-pong bi-bi	A	mixed	competitive
	BX	competitive	mixed

SUMMARY

Two-substrate enzyme-catalysed reaction may proceed by a variety of mechanisms, including the ping-pong bi-bi, compulsory-order ternary-complex, and random-order ternary-complex mechanisms. General rate equations have been derived by Alberty for two-substrate reactions which obey the Michaelis-Menten equation with respect to one-substrate at fixed concentrations of the other.

The mechanism of a two-substrate reaction may be investigated by steady-state methods, including product inhibition studies.

RECOMMENDATIONS

1. Cleland, W.W. (1970), 'Steady State Kinetics', in Boyer, P.D., *The Enzymes*, 3rd edn., Vol. 2, pp. 1-65, Academic Press.
2. Dalziel, K. (1975), 'Kinetics and Mechanism of Nicotinamide-dinucleotide-linked Dehydrogenases', in Boyer, P.D., *The Enzymes*, 3rd edn., Vol. 11, pp. 1-60, Academic Press.
3. Engel, P.C. (1977), *Enzyme Kinetics* (Chapters 5 and 6), Chapman and Hall.
4. Montgomery, R. and Sweson, C.A. (1976), *Quantitative Problems in the Biochemical Sciences*, 2nd edn. (Chapter 11), Freeman.
5. Roberts, D.V. (1977), *Enzyme Kinetics*, Cambridge University Press.
6. Tipton, K.F.(1974), Enzyme Kinetics, in Bull., A.T., Lagnado, J.R., Thomas, J.O. and Tipton, K.F., *Companion to Biochemistry* (pp. 227-52), Longman.

8 The Nature of Enzyme Catalysis

Whatever the mechanism of catalysis is, all catalysts act by reducing the energy of activation or more correctly the free energy of activation (ΔG^*) of the reaction being catalysed. Free energy is made up of an entropic and an enthalpic component. Since $\Delta G = \Delta H - T\Delta S$, hence $\Delta G^* = \Delta H^* - T\Delta S^*$, where ΔH^* is the enthalpy of activation and ΔS^* is the entropy of activation. A catalyst functions to lower the value of ΔG^* either by decreasing ΔH^* or by making ΔS^* more positive and thus helps the reaction to proceed.

In enzyme-catalysed reactions loss of entropy occurs largely in the binding steps when enzyme-substrate complexes are formed and not in the actual conversion of substrates to products. The binding of substrate molecules in close proximity to each other on the enzyme surface effectively increases their concentrations and reduces the entropy loss for the subsequent formation of a transition-state, this has been called the proximation effect.

The catalytic sites on the enzyme contribute in lowering the enthalpy by stabilizing the transition state.

8.1 MECHANISMS OF CATALYSIS

A chemical reaction is usually catalysed by one of the following mechanisms:

8.1.1 Acid-Base Catalysis

Acids can catalyse reactions by temporarily donating a proton, bases can do the same by temporarily accepting a proton. Bases may also increase reaction rate by increasing the nucleophilic character of the attacking groups. Thus, hydroxide ions displace halides from alkyl halides more rapidly than do neutral water molecules. Similarly, acids facilitate the removal of leaving groups where these are strong bases. For example, the following reaction would normally proceed extremely slowly:

$$R_3C \longrightarrow O \longrightarrow R' \longrightarrow R_3CX + {}^-OR'$$

In the presence of an acid, however, conditions are much more favourable :

$$R_3C \longrightarrow O \longrightarrow R' + H^+ \longrightarrow$$

$$R_3C \longrightarrow \overset{+}{\underset{H}{O}} \longrightarrow R' \longrightarrow R_3CX + HOR'$$

8.1.2 Electrostatic Catalysis

A transition-state may be stabilized by electrostatic interaction between its charged groups and charged groups on a catalyst. Thus, the positive charge on a carbonium ion can be stabilized by interaction with a negatively charged carboxylate ion, similarly the negative charge on an oxyanion can be stabilized by a positively charged metal ion. For example, the hydrolysis of glycine esters:

$$H_2N.CH_2C.OCH_3 + H_2O \rightleftharpoons H_2N.CH_2C.OH + CH_3OH$$

may be catalysed by cupric ions, the mechanism probably involving the following steps:

8.1.3 Covalent Catalysis

In contrast to acid-base and electrostatic catalysis where the transition-state is merely modified, covalent catalysis introduces a different reaction mechanism. In case of nucleophilic catalysis, the catalyst is more nucleophilic than the normal attacking group and so rapidly forms an intermediate which itself rapidly breaks down to give the product. For example a variety of tertiary amines catalyse the hydrolysis of esters.

In contrast, electrophilic catalysts act by withdrawing electrons from the reaction centre of an intermediate and may be termed electron sinks.

8.1.4 Enzyme Catalysis

All of the above mechanisms of catalysis are seen in enzyme-catalysed reactions. However, enzymes because of their great size and range of properties are able to impose their presence on a reaction to a far greater extent than most catalysts.

8.1.4.1 Mechanism of reactions catalysed by enzymes without cofactors

Enzymes which operate without cofactors tend to be relatively small and have relatively straightforward reaction mechanisms.

Chymotrypsin is formed by the cleavage of several peptide bonds in the inactive monomeric protein chymotrypsinogen, which is synthesized and secreted by mammalian pancreas. The active enzyme thus produced, consists of three non-identical polypeptide chains. Chymotrypsin catalyses the cleavage of peptide bonds at the carboxyl side of the amino acid (phenylalanine, tyrosine or tryptophan) residues. It also hydrolyses a variety of amides and esters and these artificial substrates have been used to investigate the enzymes in detail. Chymotrypsin-catalysed hydrolysis of an ester proceed via the formation of an acyl-enzyme. This is also true for the hydrolysis of amides and the reaction in general terms can be written as follows:

$$E + R.CO.Y \longrightarrow E.R.CO.Y \longrightarrow E.CO.R \xrightarrow{H_2O} E + RCOOH$$

Ester of amide → YH → Acyl Enzyme

8.1.4.2 Mechanism of reactions catalysed by metal-activated and metalloenzymes

More than a quarter of all known enzymes require the presence of metal atoms for full activity. Metal atoms usually exist as cations and often have more than one oxidation state as with ferrous (Fe^{2+}) and ferric (Fe^{3+}) iron. It has been noted that this positive charge can stabilize transition-states by electrostatic interactions giving one mechanism of catalysis for metals. However, irrespective of the oxidation state and charge carried, a metal ion can bind a particular number of groups (ligands) by accepting free electron pairs to form co-ordinate bonds in specific orientations.

Therefore, metal ions can be involved in enzyme catalysis in a variety of ways; they may accept or donate electrons to activate electrophiles even in neutral solution; they may act themselves as electrophiles or nucleophiles; they may mask nucleophiles to prevent unwanted side reactions; they may bring together enzyme and substrate by means of co-ordinate bonds, possibly causing strain to the substrate in the process; they may hold reacting groups in the required three dimensional orientation; they may simply stabilize a catalytically active conformation of the enzyme.

With metalloenzyme, the metal is tightly bonded and retained by the enzyme on purification. With metal-activated enzymes, the binding is less tight and the purified enzymes may have to be activated by the addition of metal ions.

It has been investigated that ternary complexes that form between enzyme (E), metal ion (M) and substrate (S) may be enzyme bridge complexes (M - E - S), substrate bridge complexes (E - S - M)

or metal bridge complexes (E - M - S). Metalloenzymes cannot form substrate bridge complexes, since the purified enzyme exists as E-M.

Oxygen atoms are often involved in the bonds of both alkali and alkaline earth metal cations, bonds of the latter being relatively stronger. The divalent cations, Ca^{2+} and Mg^{2+}, can form six co-ordinate bonds to produce octahedral complexes.

It has been shown that all possible types of ternary bridge complexes involving divalent cations can exist. Let us consider the reaction catalysed by creatine kinase:

$$Creatine + Mg\text{-}ATP \rightleftharpoons Mg\text{-}ADP + Phosphocreatine + H^+$$

The true substrate is Mg-ATP and the reaction proceeds via the formation of the complex $(Mg^{2+}\text{-}ATP)$- E - Creatine. Divalent cation binds to the α- and β- phosphates of the nucleotide but not to the terminal (γ) phosphate which is transferred to creatine. The cation helps in the orientation of the complex and may also assist in the breaking of the pyrophosphate bond by withdrawing electrons from the β- phosphate.

Superoxide dimutase is a copper-metalloenzyme which catalyses the removal of the highly reactive O^{2-} produced, for example, by oxidation of xanthine by molecular oxygen in the presence of xanthine oxidase.

$$E\text{-}Cu^{2+} + \bar{O}_2 \longrightarrow E\text{-}Cu^+ + O_2$$

$$E\text{-}Cu^+ + \bar{O}_2 \xrightarrow{+2H^+} E\text{-}Cu^{2+} + H_2O_2$$

8.1.4.3 Mechanism of reactions catalysed by enzymes requiring coenzymes

Coenzymes are organic compounds required by many enzymes for catalytic activity. They are often vitamins or their derivatives. Sometimes they can act as catalysts in the absence of enzymes but not so effectively as in conjunction with an enzyme.

As with metal-enzyme linkages, there is a range of bond strengths for coenzyme-enzyme links. Coenzymes which are tightly bound (prosthetic groups) form an integral part of the active site of an enzyme and undergo no net change as a result of acting as a catalyst. Loosely bound coenzymes can be regarded as co-substrates, since they often bind to the enzymes-protein together with the other substrates at the start of the reaction and are released in an altered form at the end of it. They are regarded as coenzymes since they usually bind to the enzyme before the other substrates are bound, since they participate in many reactions and may be reconverted to their original form by many enzymes present within cells.

Alcohol dehydrogenase needs NAD^+ as coenzyme in the catalytic oxidation of primary or secondary alcohols. NAD^+ is the first substrate to be bound and NADH is the last product to leave. The dissociation of NADH from the enzyme is the rate limiting step of the overall reaction, this is one reason why NAD^+ is regarded as a coenzyme rather than simply as a substrate. Horse liver alcohol dehydrogenase is a dimer, each sub-unit containing one binding site for NAD^+ and two sites for Zn^{2+}. Only one of these zinc ions is directly involved in catalysis. The ternary complex formed may be presented as:

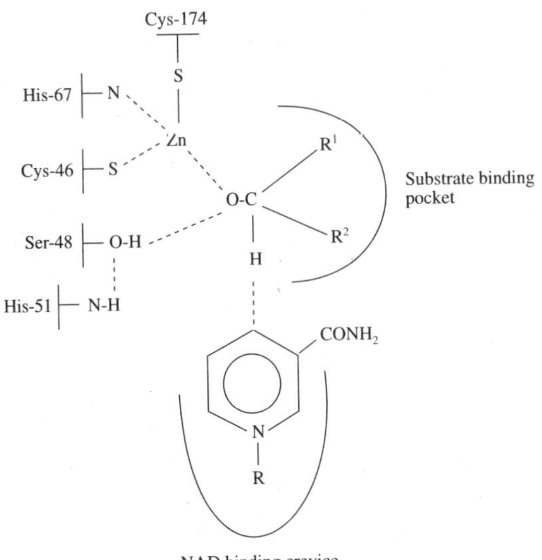

and the reaction mechanism may involve:

$$NAD^+ \quad H-C-O------Zn^{2+} \rightleftharpoons NADH \; + \quad C=O ------ Zn^{2+}$$

SUMMARY

Enzyme-catalysis can include, within a single reaction mechanism, acid-base, electrostatic and covalent catalysis as well as, proximity effects. In this way, the total free energy of activation required for the reaction to proceed is reduced and also spread out over several stages. The free energy requirement for each stage may be provided in some way from the energy made available by collisions between the enzyme-substrate complex and solute molecules.

Some enzymes can catalyse reactions without requiring cofactors. Other enzymes require the presence of metal ions for full catalytic activity. These ions can play a structural role or act as catalysts in a variety of ways. Coenzymes are also required by many enzymes. In some cases, in addition to metal ions, coenzymes usually play a catalytic role.

RECOMMENDATIONS

1. Murihead, H. (1983), Triose phosphate isomerase, pyruvate kinase and other α/β–barrel enzymes, *Trends in Biochemical Sciences*, 8, pp. 326-30.
2. Price, N.C. and Stevens, L. (1989), *Fundamentals of Enzymology*, 2nd edn. (Chapter 5), Oxford University Press.
3. Sykes, P. (1986), *A Guidebook to Mechanism in Organic Chemistry*, 6th edn., Longman.
4. Walsh, C.T. (1984), *Enzyme Mechanisms*, *Trends in Biochemical Sciences*, 9, pp. 159-62.

PART III

PART III

Investigation and Purification of Enzymes

9 Investigation of Enzymes

9.1 INVESTIGATION OF ENZYMES IN VARIOUS BIOLOGICAL PREPARATIONS

9.1.1 Choice of Preparation for the Investigation of Enzyme Characteristics

There is no ideal preparation in which the properties of an enzyme may be investigated. Different properties may have to be investigated in different preparations. Highly purified preparations are essential for the investigation of enzyme structures. They also allow enzyme characteristics (e.g. K_m and k_{cat}) to be determined and reaction mechanisms to be studied without complications arising from the presence of other enzymes and reactions. However, the removal of an enzyme from its natural environment may give a distorted impression of its characteristics *in vivo* and the elucidation of its physiological role more difficult. In order to get the complete picture, it is necessary to investigate enzymes in a purified form and also in preparations where some cellular organization has been retained.

Tissue slice technique enables *in vivo* studies to be carried out on cells which retain a considerable degree of their organization. A thin slice is cut from a tissue using a sharp implement such as a microtome. It is usually about 0.5 mm thick because the thinner one would contain too large a portion of cells damaged during the slicing and a thicker one would present problems relating to diffusion. After preparation the sliced tissue is incubated in a suitable medium at controlled pH and temperature and the uptake and release of substances of interest may be measured. Therefore, in contrast to *in vivo* studies, the metabolism being investigated is that of one particular type of tissue, e.g. liver. It is however, still difficult to relate the results to any particular enzyme, also in the absence of an intact vascular system, ingredients from the medium can only reach the innermost cells by diffusion through the others.

Problems that result from different cells of a tissue not being equally accessible to the medium may be overcome by separating the cells from each other. This is often done by treating the isolated tissue with collagenase or other proteolytic enzyme to break down the matrix and enable the individual cells to be dispersed. The conditions of proteolysis should be sufficiently mild so that each isolated cell may be retained most of its *in vivo* characteristics, but it is very difficult to prevent some damage occurring to surface proteins. Under favourable conditions, isolated cells in culture may grow and divide and thus increase the number of cells available for investigation. The technical problem in long-term culture is the prevention of contamination particularly by micro-organisms and there is also a tendency for cells to lose their precise function as time passes.

All investigations of enzymes in intact cells must be complicated by transport through the plasma membrane. A simple way to overcome this problem is to prepare a cell-free system. This may be done by homogenization of a tissue in a suitable isotonic medium (e.g. 0.25 M sucrose), the aim usually being to disrupt the cells and increase the contents without damaging the sub-cellular organelles.

Cell-free preparations are frequently produced from the micro-organisms by subjecting them to high-frequency sound waves (sonication), this procedure also causes considerable damage to cell membranes. It is important to maintain low temperatures throughout the production of *in vivo* preparations.

Cell-free preparations provide a convenient and reproducible means of investigating enzymes and metabolism in the presence of some important cellular components and interesting deviations from findings in pure enzyme preparations are sometimes revealed; for example different estimates for the K_m of glucokinase have been obtained in liver homogenates and in pure solution. However, regardless of the method by which cell-free systems are prepared, a considerable degree of cellular organization is lost in the process, hence the results obtained still cannot be taken to be exact representation of what occurs in the intact organism.

In short, it is only possible to investigate the characteristics of individual enzymes in preparations where cellular organization is totally or partially absent, so the characteristics observed are not necessarily identical to those existing *in vivo*. The differences between the preparation and the whole organism must always be borne in mind during the interpretation of results.

9.2   ENZYME ASSAY

The purpose of enzyme assay is to determine how much of a given enzyme of known characteristics is present in a tissue homogenate, fluid or partially purified preparation.

9.2.1 Enzyme Assay by Determining Catalytic Activity

By far the most convenient way to estimate the concentration of a particular enzyme in a preparation or fluid is to determine its catalytic activity, i.e. to find out how much substrate it is capable of converting to product in a given time under specified conditions. As discussed earlier, the initial velocity of an enzyme-catalysed reaction is given by:

$$v_0 = \frac{k_2[E_0][S_0]}{[S_0] + K_m}$$

where k_2 is the rate constant relating to product formation (i.e. k_{cat}). Therefore, at constant $[S_0]$,

$$v_0 \propto [E_0]$$

In other words, for any system where these assumptions are valid, the initial velocity of a reaction catalysed by an enzyme is directly proportional to the concentration of that enzyme at fixed substrate concentration. Hence, there is a linear relationship between enzyme concentration and catalytic activity. If a reaction involves more than one substrate then the concentration of each must be fixed.

The rate of an enzyme-catalysed reaction may also depend on the concentration of a cofactor (or cofactors). As with substrates, each cofactor must be present at a fixed concentration and preferably in excess, if a reliable and reproducible system for enzyme assay is to be obtained.

Another prerequisite for a successful assay system is that the reaction being catalysed should be capable of being accurately monitored, i.e. there should be a change in optical, electrical or other properties as substrate is converted to product. If this is not the case, it may be possible to follow the course of the reaction indirectly by coupling it to one where some such change does take place.

Regardless of this, the reaction should be carried out under fixed and suitable conditions of pH, ionic strength and temperature. Enzymes are often assayed at their optimal pH but this is not essential, an enzyme might not operate at its exact optimal pH *in vivo*, so there is no fundamental reason for investigating its activity at this pH *in vitro*. Also the pH optimum may vary with temperature and other factors. Nevertheless, the pH chosen for an assay system must be sufficiently near the optimum for an appreciable rate of reaction to take place, it must also be one where the enzyme is relatively stable, for enzyme stability can vary with pH. With some reversible reactions, the pH may influence the favourable direction; for example, assay involving lactate dehydrogenase are best performed in the direction of lactate production at pH 7, but in the direction of pyruvate formation at pH 10.

The presence of salts may affect enzyme-catalysed reactions by shifting the equilibrium of any of the steps involved. They may also effectively reduce the concentration of a substrate by complexing with it. Hence, enzyme assays must be performed under carefully controlled conditions of ionic strength and composition.

The choice of temperature is governed by two conflicting factors: reaction rate and enzyme stability. Specimens are usually stored at low temperatures prior to assay in order to prevent loss of enzymic activity, however, if they were assayed at the same temperatures, the rate of reaction would be extremely low, so the sensitivity of the assay would be very poor. At higher temperatures the reaction would proceed at a faster rate giving a more sensitive assay, but the stability of the enzyme would decrease.

Enzyme activity is measured as the amount of substrate lost per unit time and it should also be related in some way to the amount of specimen used for assay. In 1961, the enzyme commission of the IUB defined an Enzyme Unit (U), later to be known as an International Unit (IU), as the amount of enzyme causing loss of 1μmol substrate per minute under specified conditions. Later, in 1973, the Commission on Biochemical Nomenclature introduced the katal (kat) as the System International (SI) unit of enzyme activity: this is defined as the amount of enzyme causing loss of 1 mol substrate per second under specified conditions. Both units are in current usage.

Obviously there must be a direct relationship between the number of units of activity found and the amount of specimen assayed. Therefore, in order for the result of an assay to have any meaning, they must be expressed in terms of the amount of sample assayed and, where appropriate, extrapolated back to the original tissue.

However, even these units may not be ideal. For example, liver is likely to contain irregular deposits of glycogen and other non-protein material, so any results expressed as, say, katals per gram liver could vary according to which part of the same liver was cut off, weighed, homogenized and assayed. To minimize problems of this type, units of enzyme activity may be related to the total protein content of the sample being assayed, rather than to its weight or volume: this is termed specific activity and may be expressed as, for example, International Units per mg protein, or as katals per kg protein. In the example being considered, a liver homogenate would be prepared and an aliquot used for enzyme assay; another aliquot of the same homogenate would be analysed for its total protein content, so the total amount of protein in the sample used for enzyme assay could be calculated and related to the observed activity.

Specific activity is quite different from molar specific activity (or turnover number, k_{cat}), the latter, by either name, being used to relate the observed activity of an enzyme to its known molar concentration, regardless of any other proteins which may be present. Molar activity may be expressed as, for example, katals per mole enzyme.

9.3   INVESTIGATION AT SUB-CELLULAR LEVEL

9.3.1 Enzyme Histochemistry

The sub-cellular location of many enzymes may be revealed by microscopy, provided suitable fixation and staining procedures are followed.

Tissues are frozen to below -20°C and sliced using a cryostat (a refrigerated microtome). This procedure ensures that the tissue is rigid, essential for obtaining thin slices (about 10 μm thick) of satisfactory quality and also minimizes loss of enzyme activity. The slices are then usually fixed in formaldehyde (formalin), HCHO, to prevent any subsequent diffusion of proteins taking place. Formaldehyde brings about cross-linking between side-chain amino groups of proteins.

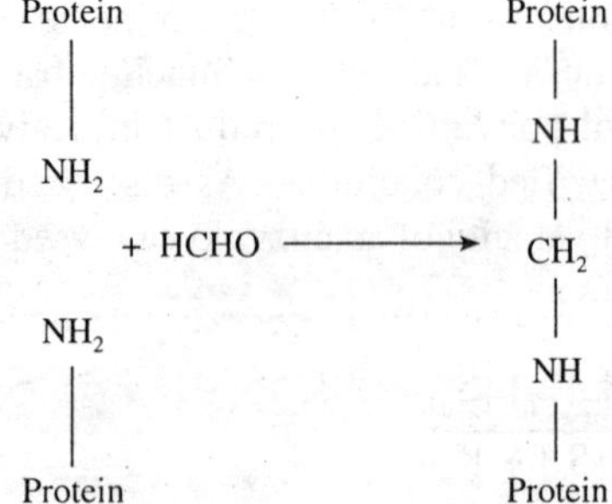

Most enzymes retain at least some activity after treatment, however some do not and these have to be investigated without prior fixation. In either case, the tissue section is treated with a buffered solution of a staining mixture containing the substrate of the enzyme under investigation. The staining process requires the conversion of this substrate to the appropriate product, and so can only take place in the regions where the enzyme is present. If possible, the product is immobilized after formation to prevent it diffusing away from the location of the enzyme.

For example, the substrate solution for the widely used lead-salt method for acid phosphatases includes sodium β-glycerophosphate and lead ions in buffer at pH 5.5. Any acid phosphatase present in the tissue hydrolyses the glycerophosphate:

$$\beta\text{-glycerophosphate} \longrightarrow \text{glycerate} + \text{Pi}$$

and the phosphate liberated reacts immediately with the lead ions to form insoluble lead phosphate ($PbPO_4$). This may be converted to a more conspicuous product, also insoluble, by treating with a solution of ammonium sulphide:

$$PbPO_4 \xrightarrow{(NH_4)_2S} PbS$$

In this way, a black precipitate of lead sulphide may be visualized wherever acid phosphatase is located.

Such investigations have revealed that acid phosphatase is characteristically associated with lysosomes.

9.3.2 The Use of Centrifugation

The sedimentation characteristics of the various sub-cellular organelles are different, so it is possible to separate them by centrifugation of a tissue homogenate, and then to investigate which enzymes are associated with each cell fraction.

The basic principle underlying is that any particle suspended in a liquid medium and being spun in a centrifuge is acted upon by a centrifugal field (G) which is determined by the angular velocity (ω) of the rotor and the radial distance (r) of the particle from the axis of rotation, according to the relationship

$$G = \omega^2 r = \left(\frac{2\pi RPM}{60}\right)^2 r$$

where RPM is the number of revolutions per minute of the rotor, ω normally being measured in radians per second.

It is convenient to express the centrifugal field as a multiple of the gravitational constant (g), this being termed the relative centrifugal field (RCF).

$$RCF = \frac{G}{g} = \left(\frac{2\pi RPM}{60}\right)^2 \frac{r}{g}$$

But $g = 980$ cm s^{-1}, provided r is measured in centimeters,

$$RCF = 1.11 \times 10^{-5} (RPM)^2 r$$

The simplest and most widely used method of separating the various sub-cellular organelles from each other is differential centrifugation. A tissue homogenate is prepared in a medium of low density (e.g. 0.25 M sucrose) and centrifuged in a series of stages, the centrifugal speed of each step being higher than for the previous one. At the end of each stage the sedimented pellet, consisting of particles of similar sedimentation characteristics, is removed.

Although heavy particles sediment faster than the lighter particles, there is originally a homogeneous distribution of particles, so some relatively light particles must be present at the bottom of the tube when centrifugation commences. Hence, each pellet must be contaminated by lighter particles which happen to be at the bottom of the tube throughout centrifugation. This problem can be overcome to some extent by resuspending each pellet in fresh homogenizing medium and repeating the centrifugation. Another problem is that all sedimented particles tend to diffuse back up the tube toward less concentrated regions particularly when the applied centrifugal field is low. A third problem with differential centrifugation is that some particles, e.g. mitochondria and lysosomes, show similar sedimentation characteristics because they are roughly the same size even though they have quite different average densities.

Centrifugation techniques where the density of the suspending medium is not uniform throughout, i.e. density-gradient centrifugation, can help to overcome some or all of these problems. A tissue homogenate or resuspended pellet in 0.25 M sucrose is layered on top of a pure solution of 0.25 M sucrose and centrifuged. The pellet formed will not be contaminated by lighter particles since no particles will be originally present at the bottom of the tube.

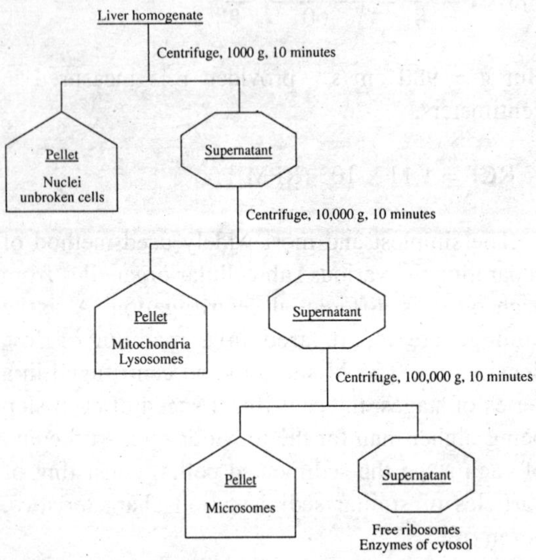

Fig. 9.1 *Simplified scheme for the fractionation by differential centrifugation of a 10% rat liver homogenate in 0.25 M sucrose at 0°C.*

Regardless of the method of separation, each fraction may then be investigated further. Enzymes still trapped in membrane-bound organelles may be liberated by techniques harsher than those used for tissue homogenization (Sec. 10.2.2). Further, centrifugation can be used to separate soluble enzymes liberated in this way from enzymes bound to the membrane of the organelle. Each fraction (or sub-fraction) may then be subjected to enzyme assay to find which enzymes are associated with it.

SUMMARY

Whole-organism studies give useful information about metabolic pathways, but can reveal little about individual enzymes. In order to investigate an enzyme in detail, it is necessary to isolate a tissue and obtain a suitable preparation. This inevitably results in a loss of cellular organization, which should be borne in mind when the results are being interpreted. Tissue preparations are kept cold (below 4°C) whenever possible to minimize autolysis and loss of enzyme activity.

Enzyme assays may be performed on tissue homogenates or other preparations. The aim is to determine the concentration (actual or relative) of the enzyme, whose characteristics must already be known. In a catalytic assay procedure, all other factors which influence the rate of reaction are made fixed and non-limiting so that the initial velocity is proportional to the concentration of enzyme present. From the rate of reaction observed and the amount of sample assayed, the activity of the enzyme preparation can be calculated. If the procedure involves the coupling of the primary reaction to a second indicator reaction, the concentration of the indicator must be made high enough to be non-limiting.

The sub-cellular compartmentation of enzymes may be investigated by enzyme histochemistry or by assay of the fractions obtained by differential centrifugation.

RECOMMENDATIONS

1. Colowick, S.P. and Kaplan, N.O. (eds), *Methods in Enzymology*, 1 (1955); Preparation and Assay of Enzymes; 70 (1980), 92 (1983), 121 (1986): Immuno-chemical Techniques, Academic Press.
2. Denton, R.M. and Pogson, C.I. (1976), *Metabolic Regulation* (Chapter 2: Practical aspects), Chapman and Hall.
3. Hopkins, C.R. (1978), *Structure and Function of Cells* (Chapters 2 and 3), Saunders.
4. Lojda, Z., Gossrau, R. and Schiebler, T. (1979), *Enzyme Histochemistry*, Springer-Verlag.
5. Ogawa, M. (1986), 'Immunoassay of circulating enzymes, *Enzyme*', 36, pp. 254-60.
6. Stryer, L. (1988), *Biochemistry*, 3rd edn. (Chapters 26, 35, 37 and 38), Freeman.
7. Wilson, K. and gouiding, K.H. (1986), *Principles and Techniques in Practical Biochemistry*, 3rd edn. (Chapters 1-4), Edward Arnold.
8. Yelton, D.E and Scharff, M.D. (1981), Monoclonal antibodies, *Annual Review of Biochemistry*, 50, pp. 657-80.

10 Extraction and Purification of Enzymes

A living organism contains literally thousands of different molecules which range from complex, high molecular weight compounds such as proteins, nucleic acids, carbohydrates and complex lipids to small molecules of organic acids, bases, sugars and salts. A large number of the proteins may be enzymatically active. The organism is precisely organized at the cellular level so that distribution of enzymes is not uniform.

10.1 SELECTION OF STARTING MATERIAL FOR EXTRACTION

While selecting the starting material for the extraction of enzyme the following points must be taken into consideration:

10.1.1 Objectives of Purification

In studying a particular type of enzyme regardless of its source, starting material chosen must contain large amounts of the enzyme and which is in plentiful supply. On the other hand, if the study is to deal with a specific enzyme from a specific source then there is no option but to start with that material regardless of concentration of the enzyme.

10.1.2 Variation of Enzyme Concentration with Age

The level of enzyme is not the same throughout the life cycle of an organism. The enzyme may be essentially absent at an early stage, rise to a maximum level and then decrease again. Certain enzymes are present in green fruit but not in ripe and vice versa. One of the problems encountered in the assay of enzyme activity in green fruits is the high level of phenolic compounds present. When cells are broken during homogenization; etc., the phenolic compounds combine with and inactivate enzymes. Thus, it is important to note that green fruit may contain high levels of certain enzymes but on the basis of the assay it might be considered that the enzyme activity is low.

10.1.3 Localization of Enzymes

Enzymes are localized in organisms. Enzymes in fruit are often concentrated near the skin and/or near the pit with low concentrations of enzymes in the bulk of the tissue. It is well worth the effort to examine the concentration of enzyme in each of the parts of an organism and to use the richest source, with the realization that an enzyme from one organ is probably not quite the same as the enzyme with the same specificity from another organ. This selection process may lead to several-fold (10 to 100) enrichment over that obtained with the whole organisms.

Within an organ there is not a uniform concentration of enzymes throughout. In a seed, a specific enzyme may be concentrated in the aleurone layer, in the endosperm or in another part of the seed. In a leaf, the enzyme may be in the grana, chloroplasts, mitochondria, etc. The liver contains different types of cells with different enzyme complements. The cell contains subcellular organelles—the lysosomes, the mitochondria, the nucleus, cell membrane, etc. Each of these organelles contain specific kinds of enzymes.

10.2 PRELIMINARY PURIFICATION

The method of preliminary purification is largely based on the localization of the enzyme to be purified.

10.2.1 Method for Enzymes in Extracellular Fluid

If the enzyme is in the extracellular fluid, the cells are removed by centrifugation and the fluid is reserved for purification. So many microbial enzymes are extracellular that it seems essential to caution that the extracellular broth should be examined for the desired enzyme before cell rupture is undertaken.

10.2.2 Method for Intracellular Enzymes

Intracellular enzymes may be considered to fall within the following groups:

1. Enzymes in true solution in the cytoplasm, are simple to extract, for any disruption of the plasma membrane enables the enzyme to pass into the surrounding medium. They remain in the 'final supernatant' after complete removal of all particulate components from the homogenate.
2. Enzymes bound to plasma membrane (insoluble lipoprotein material) can be obtained in solution only by procedures which dissociate the lipoprotein complex.
3. Enzymes associated with subcellular organelles such as with nuclei, mitochondria, microsomes, and golgi material. These enzymes may occur:
   a) in solution within a limiting intracellular membrane of an organelle, or
   b) associated with insoluble lipoprotein material.

Enzymes of the former group are also easy to liberate but the disruption of intracellular membranes often requires harsher conditions than disruption of the plasma membrane. Those of the latter can be obtained in solution only by procedures which dissociate the lipoprotein complex.

Thus, as far as the method of extraction is concerned, two major categories of intracellular enzymes are soluble (1 & 3a) and membrane-bound (2 & 3b).

Membranes, besides containing enzymes and other proteins, are made up largely of amphipathic lipids, i.e. lipids which contain hydrophilic and hydrophobic regions. The main classes of membrane lipids are phospholipids and glycolipids. In each case, the hydrophilic portion of the molecule is small compared with the hydrophobic part. An amphipathic lipid may therefore be represented as a molecule with a hydrophilic head and one or more hydrophobic tails, for example:

According to the Fluid-mosaic model of structures of membranes, proteins may be peripheral (or extrinsic), when they are bound to the surface of the membrane, or they may be integral (or intrinsic) when they are wholly or partly embedded in the lipid layers. Some evidences show that the parts of polypeptide chains embedded in membranes are rich in non-polar amino acid side chains, which may form hydrophobic bonds with the lipids, whereas those parts which stick out, thus coming into contact with water, contain more polar side chains and, in some instances, are attached to hydrophilic carbohydrate units. It seems that some of the pores through the membrane may be lined with protein.

According to this model, lipids and proteins can move laterally about the membrane, to some extent, but find much more difficulty in moving in a transverse direction, from one membrane surface to the other.

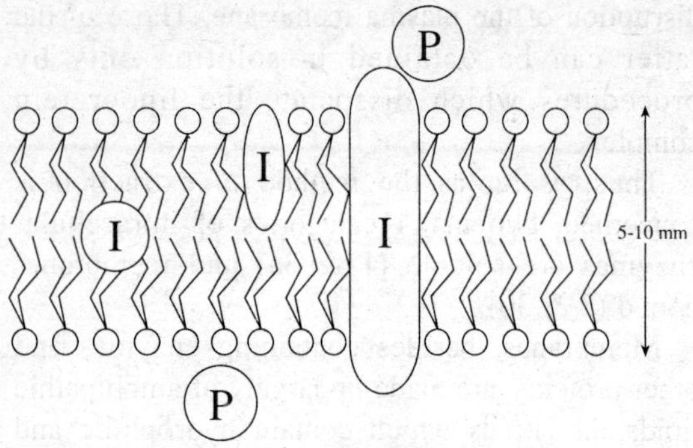

Fig. 10.1 *Fluid-mosaic model of a typical membrane (P = Peripheral protein and I = Integral protein).*

Peripheral proteins are apparently linked by electrostatic or hydrogen bonds to the polar heads of lipids or to integral proteins and can easily be dissociated from the membrane (e.g. by treating with a solution of high ionic strength, such as 1M NaCl, by freezing and thawing, or by sonication). Integral proteins, on the other hand, can only be extracted by breaking the hydrophobic interactions between lipid and protein (i.e. by dissociating a lipoprotein complex). Furthermore, for the extraction of an integral protein which is an enzyme, this must be done in such a way that enzyme activity is not lost during the process. A major problem is that an integral protein in its natural environment has some parts of its surface in contact with the hydrophobic membrane and other parts in contact with water, so structural changes are likely to take place no matter whether it is extracted into an organic or an aqueous medium.

To minimize the extraction of unwanted enzymes, cell fractionation may be carried out prior to the disruption of organelles.

10.2.2.1 Extraction of soluble enzymes

10.2.2.1a Breakage of membranes and cell walls

If the enzyme is intracellular, the first step in purification is breakage of cells and membranes and extraction of enzymes from the tissues.

Breakage of cell and organelle membranes of animal tissues is easily accomplished by homogenization in a waring blender with the extracting buffer. Rupture of plant and microbial cell walls and organelles is not so readily accomplished. Rupturing of plant cells is often facilitated by the use of acetone powder or repeated freezing and thawing.

Drying with acetone facilitates extraction of enzymes from dispersions of whole tissue or of cytoplasmic particles. The method depends on rapid removal of water. This reduces denaturation of protein by the solvent and causes concomitant disintegration of cellular structures and disruption of certain lipid-protein bonds. After complete drying and mechanical grinding to a powder, the enzymes are extracted with dilute buffer or salt solutions.

In case of repeated freezing and thawing, disintegration is associated with physical damage by ice crystals combined with plasmolysis due to the high salt concentration of the remaining fluid. Repetition of the procedure disrupts mitochondria and other cellular structures as well as whole cells.

Microbial cells may also be disrupted by grinding with sand, by homogenization in waring blender containing glass beads, by sonic oscillation, or by forcing cells through a small orifice under high pressure. Yeast cells are often broken by autolysis in the presence of toluene.

10.2.2.1b Extraction

The materials from which enzymes is to be extracted is mixed with three times the volume of extracting fluid. (Three times the volume of extracting fluid is mixed with the materials from which enzyme is to be extracted.) The suspension is stirred mechanically, with care to avoid frothing, for one to two hours. Insoluble material is removed either by centrifugation or vacuum filtration. Re-extraction of the insoluble material, using about half the initial volume of solvent, is desirable to increase recoveries.

The temperature of extraction should normally be held at about 0°C to prevent enzymic changes during extraction. Moreover, many tissues contain relatively large amounts of proteolytic enzymes. The possibility of proteolysis, including that of the enzyme under purification is very real at higher temperatures.

The choice of pH of extraction is determined by the stability of the enzyme, the nature of the enzyme complex, and the solubility of the enzyme. Electrostatic linkages may frequently be disrupted by adjustment of pH, particularly to acid values. However, since the isoelectric points of many enzymes are in the acid range, these proteins are generally more soluble at alkaline pH values. Hence, it may be advantageous to vary pH of extraction within the range of enzyme stability. The buffer is also needed to protect the enzymes from large quantities of acids released from the vacuoles on rupture of the cell. The high ionic strength helps to desorb the enzyme from its attachment to the membrane.

10.2.2.2 Extraction of membrane-bound enzymes

After breakage of cells and membranes either by use of acetone powder or repeated freezing and thawing, the most important thing for the extraction of such enzymes is the selection of most appropriate solvent for extraction. Enzymes of microsomes and of mitochondrial membranes are more or less firmly bound to lipid material. In order to separate proteins from these and other lipoprotein complexes it is necessary to remove lipid without denaturation of the associated protein.

It has been determined that the relative effectiveness of organic solvents for dissociating lipoprotein complexes in aqueous media and for releasing unsaturated protein in general may be expressed as follows:

1. Most effective: *n*- and *iso*-butanol;
2. Partially effective: *sec*-butanol, cyclohexanol, *tert*-amyl alcohol;
3. Ineffective effective: all other solvents tested, including chloroform, CCl_4, toluene, ether, acetone.

Hence butanol has a unique effect, which may be attributed to:

1. a very marked lipophilic property and special affinity for phospholipids. This results in rapid penetration of the complex.
2. Concomitant hydrophilic property, partly expressed by the solubility in water of 0.15 per cent at 0°C.
3. Orientation of the butanol molecules between the lipid and water molecules, thus producing a detergent-like action, *n*-butanol and *iso*-butanol are the only alcohols of the homologous aliphatic series possessing both hydrophilic and lipophilic properties to an appreciable extent.
4. Displacement of lipid from the protein by the butanol. Effective competitive action prevents the displaced lipids from reforming a complex with protein. The butanol possibly competes with the phospholipid for polar side chains of the protein.

Although butanol is a valuable lipid solvent when used with dried materials, it is even more effective in dissociating the lipoprotein complexes when these are in aqueous media. By contrast, for effective removal of protein-bound lipid by other organic solvents, it is necessary to remove associated water by drying the material, or by freezing to -25°C, or by addition of TCA, methanol, or other protein-denaturing agents.

10.3 REMOVAL OF NUCLEIC ACIDS

The extract from a microbial or plant source or from the nucleus, ribosomes or microsomes of animals is likely to contain large amounts of nucleic acids. These polyanionic macromolecules interfere with protein separations and should be removed at this stage. Adjustment of the pH to 5.0, or addition of basic compounds such as, protamine sulphate or streptomycin to the solution precipitates the nucleic acids.

10.4 CONCENTRATION OF ENZYME EXTRACT

The enzyme concentration in starting material is often very low. The volume of material to be processed is generally very large, and substantial amounts of waste material must be removed. Thus, if economical purification is to be achieved, the volume of starting material must be decreased by concentration. Only mild concentration procedures that do not inactivate enzymes can be employed. These procedures include either concentration by removal of water and other small molecules or precipitation method.

10.4.1 Concentration by Removal of Water and other Small Molecules

10.4.1.1 Addition of a dry matrix polymer

This is one of the simplest and quickest methods of concentrating solutions of proteins, requiring minimal apparatus. A dry inert matrix polymer, such as sephadex, is added to the protein solution and allowed to absorb the water and other small molecules. The pores within the matrix are too small to allow the protein to be absorbed. When the matrix has swollen to its full extent the remaining protein solution is removed after the matrix has been settled by gravity, filtration or centrifugation.

10.4.1.2 Ultrafiltration

In ultrafiltration, water and other small molecules are driven out through a semi-permeable membrane by a transmembrane force such as centrifugation or high pressure. For ultrafiltration, the membrane pores range in diameter from 1 to 20 nm. The diameter is chosen such that the protein of interest is too large to pass through.

10.4.1.3 Freeze drying or lyophilization

In contrast to ultrafiltration, lyophilization also results in concentration of any salt present in the initial solution, in addition, lyophilization may cause greater losses in enzyme activity. Lyophilization is, however, an invaluable method both for concentrating small molecular weight peptides which are not retained by ultrafiltration membranes, and for obtaining a dry powder of protein. Once obtained a dry powder of enzyme is more stable than an aqueous preparation of enzyme, since many degradation processes require the presence of water. Hence, many commercially available proteins are obtained as freeze-dried powders.

10.4.2 Concentration by Precipitation

As discussed earlier (Sec. 3.2.3) the solubility of a protein molecule in an aqueous solvent is determined by the distribution of charged hydrophilic and hydrophobic groups on its surface. The charged groups on the surface interact with ionic groups in the solution (Figure 10.2). Protein precipitates are formed by aggregation of the protein molecules, induced by changing pH or ionic strength, or by addition of organic miscible solvents or other inert solutes or polymers.

10.4.2.1 Precipitation with salts

High salt concentrations act on the water molecules surrounding the protein and change the electrostatic forces responsible for solubility (Sec. 3.2.3.2). Ammonium sulphate is commonly used for precipitation; hence it is an effective agent for concentrating enzymes. Enzymes can also be fractionated, to a limited extent, by using different concentrations of ammonium sulphate. In industries where enzyme is purified on a large scale, the corrosion of stainless steel and cement by ammonium sulphate is a disadvantage, and causes additional problems in waste water treatment. Sodium sulphate is more efficient from this point of view, but it is less soluble and must be used at temperatures of 35-40°C.

10.4.2.2 Precipitation with organic solvents

Organic solvents influence the solubility of enzymes by reducing the dielectric constant of the medium. The solvation effect of water molecules surrounding the enzyme is changed, the interaction of protein molecules is increased, and therefore, agglomeration and precipitation occur (Sec. 3.2.3.3). Commonly used solvents are ethanol and acetone. Satisfactory results are obtained only if the concentration of solvent and the temperature are carefully controlled because enzymes can be inactivated easily by organic solvents.

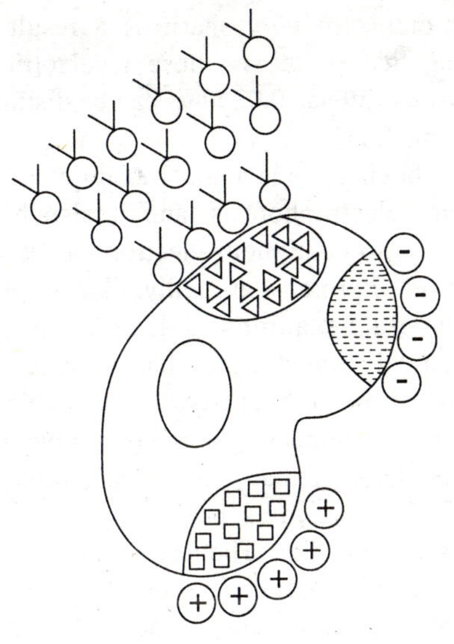

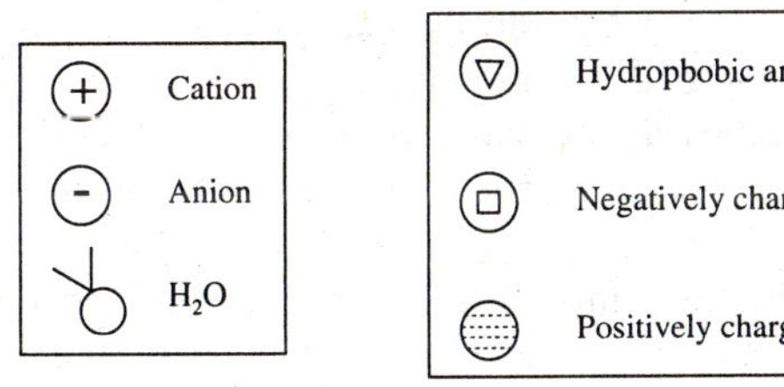

Fig. 10.2 *Schematic presentation of a protein showing negatively and positively charged areas on the protein interacting with ions in the solution. The hydrophobic areas on the protein interact with water molecules causing an ordered matrix of water molecules to form over these areas.*

10.4.2.3 Precipitation with polymers

The polymers generally used are polyethylenimines and polyethylene glycols of different molecular masses. The mechanism of this precipitation is similar to that of organic solvents and results from a change in the solvation effect of the water molecules surrounding the enzyme. Most enzymes precipitate at polymer concentrations ranging from 15 to 20 per cent.

10.4.2.4   Precipitation at the isoelectric point

Proteins are ampholytes and carry both acidic and basic groups. The solubility is markedly influenced by pH and is minimal at the isoelectric point at which the net charge is zero (Sec. 3.2.3.1). Because most proteins have isoelectric points in the acidic range, this process is also called acid precipitation.

10.5 FURTHER PURIFICATION

The crude extract, partially purified may be extracted in a variety of ways to increase the purification of the relevant enzyme.

10.5.1 Electrophoresis

Electrophoresis is mainly an analytical procedure, since it is ideally suited to the separation of small amounts of material, but it has also been used for the purification of proteins. The rate and direction of migration of a protein in an electric field depends on its net-charge at the pH used and also on the size of the molecule, since this imposes some resistance to movement.

Electrophoresis in various forms is a very effective method of separation. In zone electrophoresis, a mixture of proteins is introduced at a common point (the origin) and allowed to move in an electric field, usually in a horizontal direction, each molecule traveling in the same zone as others of the same charge and size. To minimize diffusion, the whole process is usually carried out in a solid, but porous, support medium (e.g. starch gel) through which the buffer and proteins permeate. When separation has been achieved, as indicated by cutting a narrow longitudinal section from the support medium and staining for proteins (e.g. with Amido Black), then the rest of the support medium may be cut into strips across the direction of travel and contents of each extracted by elution.

In discontinuous (disc) electrophoresis, electrodes are placed in separate buffer reservoirs linked vertically by gel (usually polyacrylamide) supported by glass plates or cylinders. The discontinuity is in the gels and buffers used, pH differences being used to concentrate the sample into a very narrow band as it passes through a large-pore spacer gel on its way to the running gel where the actual separation of the sample components takes place (Figure 10.3). This running gel has smaller pores than the spacer gel, and separation is achieved by a combination of electrophoresis and gel filtration. On completion of the electrophoresis and after staining for proteins, it is easy to see if more than one band is present, for each is extremely narrow and well-defined.

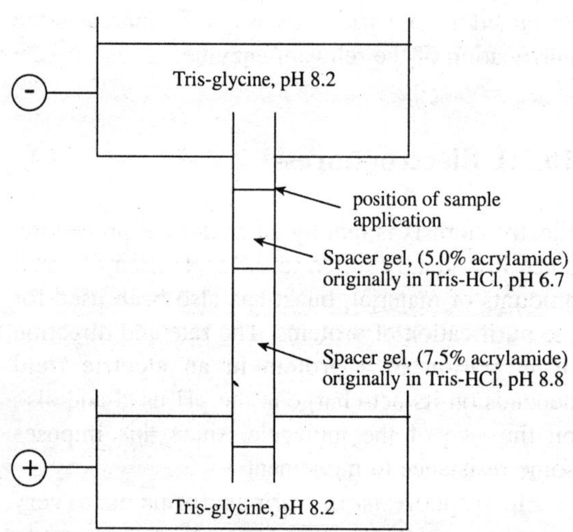

Fig. 10.3 *Apparatus set up for discontinuous (disc) electrophoresis of proteins.*

Another relevant technique is isoelectric focusing, which is an example of moving boundary electrophoresis rather than zone electrophoresis. A pH gradient is set up between the electrodes by allowing an acid (e.g. phosphoric acid) to diffuse in from the anode end and a base (e.g. ethanolamine) from the cathode end. This can be stabilized by introducing a mixture of ampholytes (e.g. synthetic and natural amino acids) selected so that their individual isoelectric points cover the required pH range. Each ampholyte will move in the electric field to the position in the pH gradient corresponding to its isoelectric point and then remain there, since it will no longer have a net charge. The protein sample is then introduced and electrophoresis continued, possibly for several days, until each component reaches a stationary position at its isoelectric pH. The zones containing each protein are very sharp as a result of this focusing and proteins whose isoelectric points differ by as little as 0.02 units can be distinguished by this method.

Isotachophoresis is another example of moving boundary electrophoresis, and it also has some similarities to continuous electrophoresis. For the separation of proteins at mildly alkaline pH, when most would be anions, a leading anion (e.g. phosphate) is added which has a faster mobility towards the anode than any of the sample anions. A trailing or terminating anion (e.g. glycine), with a slower mobility than any in the sample, is also added. The system is buffered by a common-counter ion, e.g. Tris. The leading and terminating anions are applied at different sides of the sample, the leading anion nearest the anode. A high voltage is applied and, when steady-state is achieved, all components will migrate towards the anode in discrete zones at the same velocity ('*iso tacho*', Greek for 'same speed'), the sample anions arranged in order of their mobilities. Spacers similar to the ampholytes used in isoelectric focusing may be added to separate the zones of sample anions.

10.5.2 Chromatography

Various chromatographic techniques are used to purify proteins.

Cellulosic ion exchange chromatography has proved of more general application to the separation of proteins. The resins, most frequently used, are diethylaminoethyl (DEAE)-cellulose which is an anion exchanger, and carboxymethyl (CM) cellulose, a cation exchanger. The DEAE group, $OC_2H_5N^+H(C_2H_5)_2$, is highly positively charged at pH 6-8, so DEAE-cellulose is most useful for the chromatography of proteins which are negatively charged in this pH range. Similarly CM-cellulose, cellulose $-OCH_2COO^-$, is most applicable for the separation of proteins which are positively charged at around pH 4.5.

Elution of proteins from the columns may be brought about by changes in either salt concentration or pH; as the concentration of salt (e.g. NaCl) increases, protein is displaced from DEAE-cellulose by the anion (Cl^-) and from CM-cellulose by the cation (Na^+). If the pH is altered over the relatively narrow working range, proteins are eluted as their isoelectric point is reached, since they then have no net charge with which to bind to the resin.

with swollen gels which separate components of a sample on the basis of molecular size: molecules too large to enter the pores of the gels will pass quickly through the column; smaller molecules will pass through the column more slowly, the actual speed for each component being dependent on the ease with which its molecules can pass into the gels and be thus retarded. Gels commonly used, include cross-linked dextrans (Sephadex), cross-linked agarose (Sepharose), and cross-linked polyacrylamide (Biogel). These are graded to pore size, and thus to the size (and hence molecular weight) of protein molecule which can enter. For a successful separation, the enzyme of interest must not be too large to penetrate the gel, because it would then pass straight through the column together with all other proteins of similar and greater molecular size.

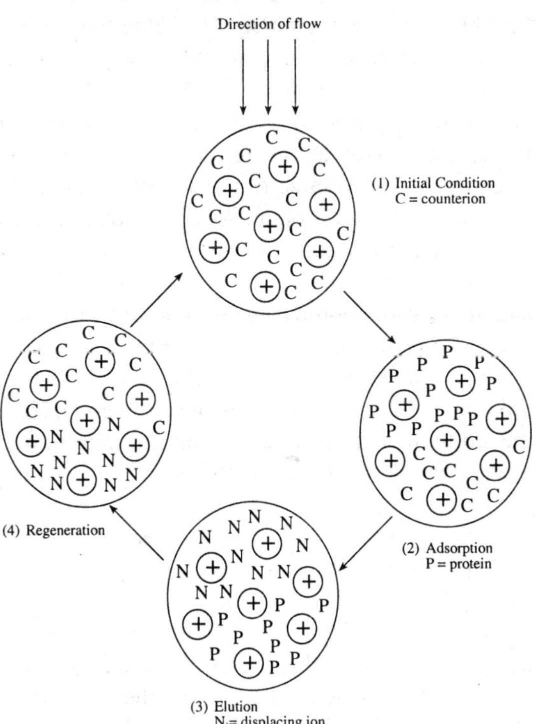

Fig. 10.4 *Stages in the purification of a protein (P) using anion exchange chromatography.*

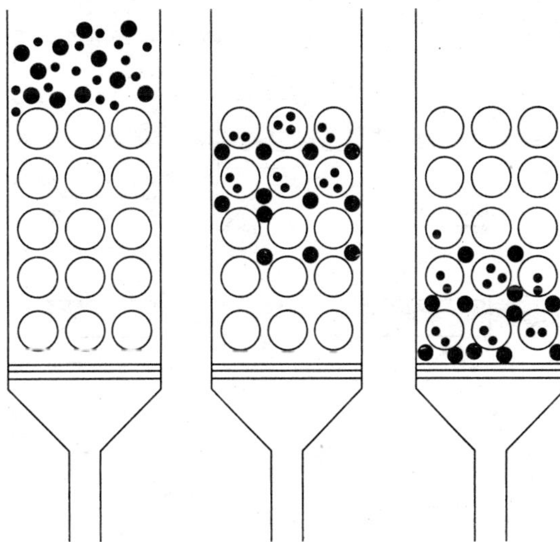

Fig. 10.5 *Stages in a chromatographic separation on a gel-filtration column.*

Gel filtration (molecular-sieve chromatography), another important technique in enzyme purification, is carried out with columns packed

Affinity chromatography is a bio-specific process and thus ideally suited for the separation of one protein from all others, including those which resemble it so closely in physical characteristics that separation by any other procedure is extremely difficult. The column is packed with an inert matrix (e.g. agarose) to which ligands for the required enzyme have been attached. These immobilized ligands may be, for example, substrate analogues, and must be covalently linked to the matrix in such a way that they can still bind to their enzymes, if this is present. The protein mixture is applied to the column, and the relevant enzyme is trapped by the immobilized ligands, while all other proteins pass through and are discarded. The enzyme is then liberated from the column either by eluting with a deforming buffer at a pH which changes the characteristics of the enzyme and no longer allows it to bind to the immobilized ligand, or by the use of a competitive counter-ligand, which displaces the immobilized ligand on the enzyme. In both cases, the enzyme passes through the column and can be collected, now free of other proteins (Figure 10.6).

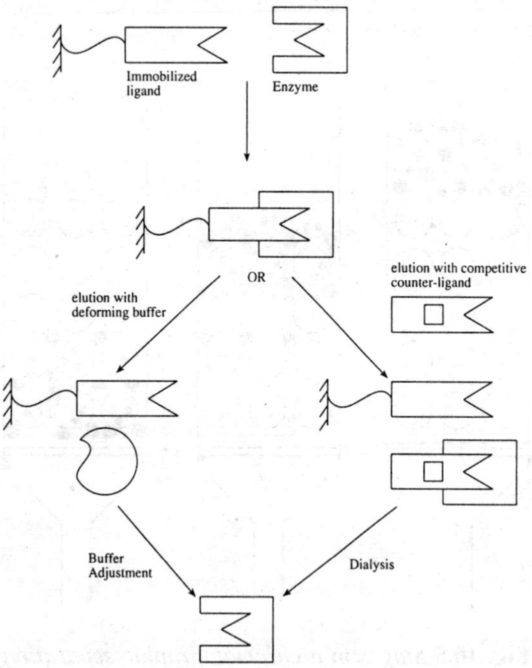

Fig. 10.6 *Representation of the process involved in affinity chromatography.*

10.5.3 Criteria of Purity

At each stage of purification procedure, an assay for the enzyme being purified should be performed on all fractions and its specific activity in each fraction determined. This helps to know which fractions contain the important enzyme and enables the degree of purification to be calculated.

With each successive purification step, the specific activity of the fraction (or fractions) containing the enzyme should be greater than before until complete purification is achieved and the specific activity reaches a limiting value. However, the finding of the same specific activity value before and after a purification step does not necessarily mean that the enzyme preparation is completely pure. It could simply mean that contaminating proteins have passed through the procedure in the same fraction as the enzyme.

10.6 DETERMINATION OF MOLECULAR WEIGHT

When an enzyme has been purified, its molecular weight may be determined. Gel filtration, which separates molecules on the basis of size (Sec. 10.5.1), provides one way of doing this. A packed column is calibrated by applying proteins of known molecular weight to the top of the column and determining the volume of the buffer required for the elution of each protein. As each protein leaves the column there should be an absorbance peak at 280 nm in the column eluate, thus providing a simple way of monitoring the elution. From the data obtained, a graph may be drawn of elution volume against molecular weight. The protein of unknown molecular weight is then passed through the column and its elution volume determined exactly as for the marker proteins.

A technique which is commonly used to determine molecular weights of proteins and which does not assume that the molecules are spherical is SDS-gel electrophoresis. SDS (sodium dodecyl sulphate, $CH_3(CH_2)_{11}SO_3^- Na^+$) is an anionic detergent whose hydrocarbon chain can become linked to hydrophobic regions in the interior of globular proteins, leaving the ionized sulphonate ($-SO_3^-$) group jutting out into the surrounding medium. Many SDS molecules can bind in this way to a single protein molecule, disrupting its natural shape and giving it a large negative charge, regardless of the net charge originally possessed by the protein. Large protein molecules will complex with more SDS than small ones, so all will have roughly the same charge/mass ratio, and all will have much the same shape. Therefore, if molecules of different proteins are complexed with SDS and subjected to zone electrophoresis in an unrestrictive medium, all should travel together, regardless of their original characteristics. However, if electrophoresis is performed in a support medium with a restrictive pore size (e.g. polyacrylamide gel), the complexes from each of the different proteins will be separated, entirely on the basis of size. Since the size of each protein-SDS complex must be dependent on the original size of the protein molecule, a system can be calibrated by the use of marker proteins of known molecular weight, hence the molecular weight of a test protein may be determined.

SUMMARY

Soluble enzymes may be extracted from cells simply by the disruption of membranes, the actual technique employed depending on the nature of the cell, the intracellular location of the enzyme and the stability of the enzyme. In each case, the pH, salt and organic solvent concentration of the medium must be carefully chosen to ensure that the enzyme being extracted remains in solution while the cell debris is removed by centrifugation.

Special procedures have to be employed to extract enzymes which form an integral part of membranes (e.g. extraction with a detergent), otherwise they would be discarded with the cell debris.

Techniques for the purification of extracted enzyme include electrophoresis, ion exchange chromatography, gel filtration and affinity chromatography. Enzyme assay should be performed after each purification step, and the specific activity of each fraction determined. As purification proceeds, the specific activity of the enzyme preparation should rise to a limiting value.

After purification, the molecular weight of the enzyme may be determined by gel filtration or SDS-gel electrophoresis.

RECOMMENDATIONS

1. Chapman, D. (1983), *Biomembrane Structure and Function*, Macmillan.
2. Colowick, S.P. and Kaplan, N.O. (eds.), *Methods in Enzymology*, 1 (1955); *Preparation and assay of enzymes*, 22 (1971); *Enzyme purification*, 34 (1974); *Affinity chromatography*, 104 (1984); *Enzyme purification*, Academic Press.
3. Kerese, I. (1984), *Methods of Protein Analysis*, Ellis Horwood.
4. Price, N.C. and Stevens, L. (1989), *Fundamentals of Enzymology*, 2nd edn. (Chapters 2 and 8), Oxford University Press.
5. Scawen, M.D. and Melling, J. (1985), 'Large-scale Extraction and Purification of Enzymes', in Wiseman, A., *Handbook of Enzyme Biochemistry*, 2nd edn., Ellis Horwood.
6. Smith, R.L. and Oldfield, E. (1984), 'Dynamic Structure of Membranes by Deuterium NMR', *Science*, 225, pp. 280-8.
7. Sofer, G.K. and Nystrom, L.E. (1989), *Process Chromatography*, Academic Press.
8. Wilson, K. and Gouiding, K.H. (1986), *Principles and Techniques in Practical Biochemistry*, 3rd edn. (Chapters 1-4, 6 and 7), Edward Arnold.

ENZYMES IN FOOD

SECTION B

11 Importance of Enzymes in Food

The activity of the enzymes in food is a field of enzymology that has been studied for a long time. It first contributed to the estimation of the degree of food freshness, the evaluation of the effectiveness of thermal processing, the study of maturation, and the detection of the onset of senescence.

Food enzymology has gone far beyond these applications. In this section, the current significance of enzymes in food will be discussed, and a detailed inventory of the essential enzymes, their characteristics and applications in some group of foods will be made.

11.1 ORIGIN OF ENZYMES

11.1.1 Native Enzymes

Enzymes can be normal constituents in food and products in food formulations. Native enzymes do not necessarily show activity in these foods for various reasons, including absence of substrates, localization of enzymes making the substrates inaccessible, unfavourable physiological conditions, presence of inhibitors, etc. Many milk enzymes, present in constant quantity, do not find enough substrate in their medium. Polyphenol oxidases, which are responsible for enzymatic browning, normally do not occur in fruits and vegetables with intact tissue structure, despite the presence of substrate; enzymes and substrates are often localized in different compartments.

11.1.2 Contaminating Enzymes

Some enzymes may be introduced to a food by various microbial contaminants, or during disease or special physiological conditions that occur in the producing organism. The presence of large quantities of reductases in milk is the indication of a bacterial contamination. The presence of catalase in this medium can have two main causes: the enzyme may arise from leucocytes and indicate the presence of blood and colostrum, which indicates mammary infection or it may be caused by microbial contamination.

11.1.3 Controlled Addition

The enzymes may be introduced voluntarily from more or less purified preparations, or even during complex fermentations by selected yeasts. The clarification of fruit juices by adding pectin esterases and fungal polygalacturonases is an example, just as the addition of flavorase exogenous enzymes to fruits and vegetables that have been blanched. Blanching destroys the native enzymes without greatly changing the more stable precursors of flavour. The addition of exogenous enzymes can partially give back the flavour to the fresh product formed from the native precursors. The following is a list of important industrial food enzymes and their applications. The majority of them are hydrolases and oxidoreductases.

1. α-Amylase

(a) Conversion of starch to dextrins (liquefaction) in the production of corn syrup.
(b) Supplement to flour types low in α-amylase to ensure a continuous supply of fermentable sugar to yeast growth and gas production in dough making.
(c) Solubilization of adjuncts (non-malt carbohydrate materials from barley) used in brewing.

2. Glucoamylase

(a) Conversion of dextrins to glucose in the production of corn syrup.

(b) Conversion of residual dextrins to fermentable sugar in brewing for the production of light beer.

3. β-Amylase

Production of high-maltose syrup.

4. Xylose (glucose) isomerase

Isomerization of glucose to fructose in the production of high-fructose corn syrup.

5. β-Glucanase

Breakdown of β-glucans in malt and other raw materials to aid filtration of wort after mashing in brewing.

6. Lipase

(a) Enhancing flavour development and shortening the time for cheese ripening.
(b) Production of specialty fats with improved qualities.

7. Papain

(a) Used as meat tenderizer.
(b) Used in brewing to prevent chill-haze formation by digesting the proteins that can otherwise react with tannic substances to form insoluble colloid particles.

8. Pectinase

Treatment of fruit pulp to facilitate juice extraction and for clarification and filtration of juice.

9. Lactase

(a) Additive for dairy products for individuals lacking lactase.
(b) Breakdown of lactose in whey products for manufacturing polylactide.

10. Glucose oxidase

(a) Conversion of glucose to gluconic acid to prevent Maillard reaction in egg products caused by high heating used in dehydration.
(b) Potential to remove O_2 in food packing for protection against oxidative deterioration.

11.2 AIMS OF ANALYSING ENZYMATIC ACTIVITY

11.2.1 Monitoring Heat Treatments

The inactivation of enzymes and the destruction of micro-organisms by heat follow the same laws and are very often governed by first-order reactions.

Under physicochemical treatment conditions, certain enzymes chosen from among the normal constituents of the sample food that are present at a constant rate show normal thermal sensitivities similar to those of the most thermoresistant groups of micro-organisms that are targeted for destruction. Detection of the total inactivation of these enzymes during the treatment (pasteurization, sterilization) will indicate, whether it was effective or the entire food has been processed. Tests conducted in this manner are simple, quick, and more convenient than those counting viable micro-organisms. This is well illustrated in the measurement of alkaline phosphatase and of the peroxidase activity in milk.

Some thermal processing is aimed at the destruction of native enzymes that deteriorate the food at certain stages of maturation. This is what happens when fruits and vegetables are blanched by steam or by immersion in hot water. In principle, this treatment must ensure the destruction of the most thermoresistant deteriorating enzyme.

11.2.2 Testing Hygienic Quality

The rapid determination of the activity of certain enzymes produced by micro-organisms in fresh food gives an indication of the food's degree of contamination. This is the case for the milk

TABLE 11.1

Determination of a few food components or additives that use enzymatic methods

Components	Enzymes Used	Measured Quantities
	ORGANIC ACIDS	
Acetic acid	Acetyl CoA synthetase + citrate synthetase + malate dehydrogenase	$NADH + H^+$
Ascorbic acid	Ascorbate oxidase	MIT-Formazan
Citric acid	Citrate lyase + malate dehydrogenase + lactate dehydrogenase	$NADH + H^+$
Formic acid	Formiate dehydrogenase	$NADH + H^+$
D-L-isocitric acid	Isocitrate dehydrogenase	$NADPH + H^+$
D-L-lactic acid	D-lactate dehydrogenase + glutamate pyruvate transminase	$NADH + H^+$
L-malic acid	L-malate dehydrogenase + glutamate oxaloacetate transminase	$NADH + H^+$
Oxalic acid	Oxalate decarboxylase + formiate dehydrogenase	$NADH + H^+$
	CARBOHYDRATES	
Starch	Aminoglucosidase + hexokinase + glucose-6-P-dehydrogenase	$NADH + H^+$
Fructose	Hexokinase + phosphoglucose isomerase + glucose -6-P-dehydrogenase	$NADPH + H^+$
Glucose	Glucose oxidase + peroxidase	oxidized chromogen
Lactose / Galactose	β-galactosidase + galactose dehydrogenase	$NADPH + H^+$
Maltose / Saccharose	α-glucosidase followed by glucose determination	$NADH + H^+$
Raffinose	α-glucosidase + galactose dehydrogenase	$NADH + H^+$
Sorbitol / Xylitol	Sorbitol dehydrogenase	$NADH + H^+$
	AMINO ACIDS	
L-aspartic acid	Glutamate-oxaloacetate transminase + malate dehydrogenase	$NADH + H^+$
L-asparagine	Asparaginase	
L-glutamic acid	Glutamate dehydrogenase + diaphorase	formazan (492 nm)
	LIPIDS	
Cholesterol	Cholesterol oxidase + catalase	lutidine
Glycerol	glycerokinase + pyruvate kinase + lactate dehydrogenase	$NADH + H^+$
Triglycerides	Lipase esterase + glycerokinase + pyruvate kinase + lactate dehydrogenase	$NADH + H^+$

bacterial reductases already indicated, but it is obviously not a specific method. The detection of catalase in milk associated with the measurement of acidity enables the detection of bacterial contamination or an alteration caused by leucocytes (colostrum, presence of blood, mammary infection). The presence of catalase that is linked to acidification is the index of lactic fermentation. Without acidification, it expresses a non-bacterial alteration.

11.2.3 Determination of Constituents

The enzymatic food analysis methods are numerous. By using purified enzymatic preparations from different sources, they permit the determination of constituents that are often difficult to determine and differentiate by purely chemical methods. Table 11.1 indicates a few systems used for this purpose.

11.2.4 Biodynamics of Food

The analysis of various enzymatic activities during the storage of food under diverse conditions can be used to obtain valuable information on the range of metabolic pathways that come into play in the processes of maturation and senescence. Knowledge of food biodynamics allows them to be better controlled, and contributes to an improvement in the preservation conditions.

The storage of potatoes is such an example. At a high temperature, the enzyme for soluble sugar-starch interconversion operate in favour of starch synthesis. The small quantity of saccharides present is consumed by respiration, which provides the energy necessary for synthesis. At low temperatures, it appears that phosphorylase is activated. The activity of respiratory enzymes is thus low, and there is sweetening through accumulation of soluble sugars, particularly glucose, as well as fructose.

11.3 FUTURE APPLICATIONS OF FOOD ENZYMES

11.3.1 Tailoring Enzyme Properties and Functions

Many ongoing investigations are directed to modify individual enzymes for specific functional properties. Enzymes used in food processing can be tailored to increase the efficiency of the process, and ultimately lower the cost of operations. Two key enzymes involved in corn syrup production can be improved, for example, to benefit the industry and the consumers. For complete liquefaction, starch granules need to be gelatinized by heating to about 105°C for α-amylase to act. All α-amylases used today require Ca^{++} ion for heat stability. In the following step, glucoamylase is used to further breakdown dextrins to glucose. Because glucoamylase has an acidic pH optimum and relatively low heat stability, the syrup from the liquefaction step must be cooled to about 60°C, and the pH adjusted to ~ 4.5 before saccharification. It is therefore, highly desirable to have a glucoamylase that is heat-stable at the temperature range used in liquefaction and a higher pH optimum, so that both steps can be carried out in one process. Furthermore, glucoamylase catalyses the hydrolysis of α-1,4 about 30-50 times faster than the branching (α-1,6) linkage. The yield of conversion in saccharification is ~ 96 per cent. It is desirable to add a debranching enzyme, or alternatively, glucoamylase may be modified to enhance its action on (α-1,6) bonds.

For the production of high-fructose corn syrup, the glucose syrup feedstock needs to have the Ca^{++} ions removed. For glucose isomerase, Mg^{++} is added and a pH > 7.0 is required for optimum activity. Reaction in a high-glucose medium at an alkaline pH causes glycosylation and eventually inactivation of the enzyme. Intensive effort is now focused on shifting the pH optimum of the enzyme to neutral or slightly acidic, and changing the metal preference of the enzyme. In addition, the catalytic efficiency of glucose isomerase in the conversion of glucose to fructose is lower than that in the transformation of xylose to xylulose. The former reaction utilized in food processing is not the natural physiological function in micro-organisms. There is great interest in modifying the enzyme to enhance its substrate specificity for glucose.

Lipases are used in the food oil industry for transesterification of inexpensive oils (e.g. palm oil) to produce substitutes for cocoa butter, which is a major ingredient in chocolate and confectionery manufacturing. The characteristic melting property (a low and sharp melting point at 30-40°C) of cocoa butter is derived from the positions and compositions of the fatty acids in its triglyceride structure. Palm oil with most of its 1,3-position occupied by palmitic acid can be transesterified with stearic acid to produce a cocoa butter-like replacement, using a 1,3 specific lipase. Both the stability as well as the catalytic efficiency of lipases under commercial oil processing conditions need to be improved.

Lipase is also utilized in the production of enzyme-modified cheese/butter-fat. Direct development of selected cheese flavour is achieved by lipase-catalysed hydrolysis of cheese curd or butterfat to generate free fatty acids, mostly short chains. Generation of specific flavour is dependent upon the specificity and property of the lipase employed. There has been continuous effort to screen for enzymes suitable for use in flavour development. Proteases can be utilized in this regard, but a major drawback is the requirement of precise control of hydrolysis in preventing the formation of bitter peptides. The addition of lipases and proteases to accelerate the cheese-ripening process is of economic significance. Again, the delicate control of enzyme action in the development of a proper flavour profile for the product remains a challenge to food scientists.

All these cases point to the need and potential in the utilization of enzymes in food processing. In nearly all the industrial enzymes currently used, there are properties awaiting improvements. Traditionally, the primary focus is on the screening of micro-organisms for enzymes with desired or useful properties.

11.3.2 *In Vivo* **Modification of Food Quality**

Initial interests in the application of biotechnology in agriculture have generally been concentrated in controlling pests and diseases, and in developing pesticide/herbicide tolerance in crops. Increasing efforts are now directed to the improvement of food quality. A well-known case is the production of a genetically engineered tomato with the polygalacturonase gene silenced by antisense RNA. This process represents an example of *in vivo* manipulation of an endogenous enzyme to effect alteration in the quality of a food. The success of this type of research and development substantiates the potential use of modification of enzymatic pathways in plants for:

1. Improving functional properties of proteins and other food components more suitable for food formulation and processing.

2. Tailoring the chemical composition and physical properties of food components for value-added products. For example, a potato with a high starch content and less moisture would absorb less oil on frying. The canola plant can be modified to produce oils of specifications, for example, those that are suitable for margarine. The modified oil thus eliminates hydrogenation and transesterification, and hence problem of *cis-trans* isomerization of fatty acids in margarine production.

3. Removing or reducing toxicants, or undesirable compounds in food crops to improve their nutritional value.

The field of enzymology is expanding with new information on enzyme structures and mechanisms accumulating at a rapid pace. It is the synergistic activity of both basic and applied technology that will eventually benefit the general public. Fundamental knowledge on enzyme structures and mechanisms is important for effective utilization and improvement of existing enzymes and the development of new enzymes. A thorough understanding of enzyme reactions is critical to future success in the utilization of agricultural resources and in the production of better foods.

SUMMARY

Enzymes in food may have different origins. Native enzymes can be normal constituents in food and products in food formulations. Native enzymes do not necessarily show activity in these foods for various reasons, including absence of substrates, localization of enzymes making the substrates inaccessible, unfavourable physiological conditions, presence of inhibitors. Contaminating enzymes may be introduced to a food by various microbial contaminants, or during disease or special physiological conditions that occur in the producing organism. The applications of food enzymology include, not only the estimation of the degree of food freshness, but the evaluation of the effectiveness of thermal processing, the study of maturation and the detection of the onset of senescence. Food enzymology has gone far beyond these applications.

Additive enzymes, isolated from different sources, are now introduced voluntarily to various food and food products during processing to achieve different purposes. The clarification of fruit juices by adding pectin esterases and fungal polygalacturonases is an example. The important industrial food enzymes include hydrolases and oxidoreductases.

Aims of analysing enzymatic activity in food include: (1) Monitoring heat treatments, (2) Testing hygienic quality, (3) Determination of constituents, and (4) Biodynamics of foods.

Enzymes used in food processing can be tailored to increase the efficiency of the process and ultimately lower the cost of operations. Increasing efforts are now directed to the *in vivo* manipulation of an endogenous enzyme to effect alteration in the quality of a food. The success of this type of research and development substantiates the potential use of modification of enzymatic pathways in plants for improving functional properties of proteins and other food components more suitable for food formulation and processing.

RECOMMENDATIONS

1. Barman, T.E. (1974), *Enzyme Handbook*, Springer Verlag, Berlin.
2. Erickson, D. (1992), 'Hot Potato', *Scientific American*, 267 (3), pp. 160-61.
3. Grufferty, M.B., Fox, P.F., (1988), 'Milk Alkaline Proteinase', *J. Dairy Res.*, 55, pp. 609-30.
4. Humbert, G., Alais, C. (1979), 'The Milk Proteinase System', *J. Dairy Res.*, 46, pp. 559-71.
5. IUB (1984), *Enzyme Nomenclature*, p. 647 Academic Press, New York.
6. Lelen, K. (1992), 'Ag-biotechnology Companies Move Forward on Heels of the FDA Statement on Biofoods', *Genetic Engineering News*, 12 (11), pp. 21-2.
7. Linden, G. (1986), Biochemical Aspects, Monograph on Pasteurized Milk. FIL/IDF doc. 200, Chap. V, pp. 17-21.
8. MacLeod, A. (1979), 'The Physiology of Malting', *Brewing Science*, Pollock J.R. A., 1., pp. 145-232, Academic Press, New York.
9. Pintauro, N.D. (1979), 'Food Processing Enzymes', *Recent Developments*, Noyes Data Corporation, New Jersey.
10. Skillicorn, A. (1994), 'Oilseeds Get a Genetic Makeover', *Food Processing*, 55(2), pp. 48-52.

PART I

Oxidoreductases

The oxidoreductases catalyse the oxidation or reduction of substrates. Oxidation or reduction of a substrate can occur in a number of ways as shown in equations 1 to 7 where the distinction is made on the basis of electron acceptor (B, O_2 or H_2O_2), electron donor (A or AH_2) and products formed .

$$1. \quad AH_2 + B \longrightarrow A + BH_2$$

$$2. \quad AH_2 + O_2 \longrightarrow A + H_2O_2$$

$$3. \quad 2AH_2 + O_2 \longrightarrow 2A + 2H_2O$$

$$4. \quad A + H_2O + B \longrightarrow AO + BH_2$$

$$5. \quad A + H_2O_2 \longrightarrow AO + H_2O$$

$$6. \quad A + O_2 \longrightarrow AO_2$$

$$7. \quad A + O_2 + BH_2 \longrightarrow AO + B + H_2O$$

By far the largest number of oxidoreductases (Table 1) belong to type 1 reactions in which enzymes catalyse oxidation of the substrate by removal of hydrogens and/or electrons through participation of an acceptor B.

The Oxidoreductases

Type of reaction	Examples of enzymes
1	Alcohol dehydrogenase
2	Glucose oxidase, Galactose oxidase
3	Polyphenol oxidase, *o*-diphenoloxidase
4	Xanthine oxidase
5	Catalase, Peroxidase
6	Lipoxygenase
7	Polyphenol oxidase (Hydroxylation)

12 Glucose Oxidase

Glucose oxidase (β-D-glucose: oxygen 1-oxidore-ductase, E.C.1.1.3.4) catalyses the oxidation of β-D-glucose to δ-D-glocuno-1,5-lactone coupled with the reduction of O_2 to H_2O_2.

The enzyme is found in a number of fungal sources, including *Aspergillus oryzae*, *Penicillium notatum*, *Penicillium glaucum*, and *Talaromyces flavins*, most studies have been done on the enzyme purified from *Aspergillus niger*.

The enzyme from different fungi share certain similarities in that all are dimers composed of two identical subunits, have 2 moles of firmly bound FAD per mole of protein, and contain 11-20 per cent carbohydrates.

Each enzyme subunit contains one disulphide bond and one free sulfhydryl. The two subunits are non-covalently associated; only the dimer form is active. ^{31}P-NMR shows a covalent phosphorus residue located near the surface of the enzyme molecule. The disubstituted phosphorus forms a phosphodiester linkage between two amino acid residues similar to sulphide links stabilizing proteins.

12.1 REACTION MECHANISM

According to the proposed mechanism glucose oxidase does not oxidize β-D-glucose by direct combination of molecular oxygen with the β-D-glucose. Evidence for this include the following observations:

1. Glucose oxidase is not inhibited by HCN and CO, indicating that it does not contain an essential metal ion. In general, oxidases contain metal ions as essential cofactors.
2. They are able to utilize such compounds as thionine, methylene blue, and quinone (hydrogen acceptors) and can replace O_2. The ability to utilize such compounds is characteristic of dehydrogenases but not of oxidases.
3. Narcotics inhibit the oxidative capacity of the enzyme which is in agreement with that observed for other dehydrogenases but not with oxidases.
4. The high substrate specificity of the enzyme is a general property of dehydrogenases but not of oxidases.
5. There is a direct correlation between riboflavin content and enzymatic activity. The highly purified enzyme is yellow and contains two molecules of FAD per molecule enzyme. Removal of FAD leads to complete loss of activity which can be restored on addition of FAD.
6. Glucose oxidase in the oxidized form has an absorbance maximum at 280 nm, typical of all proteins, and absorbance maxima at 377 and 455 nm due to FAD. On addition of glucose under anaerobic conditions the absorbance maxima at 377 and 455 nm disappear. When oxygen is admitted to the system, the two absorbance maxima reappear.
7. Glucose oxidase does not catalyse the complete reaction under anaerobic conditions and the reaction rate is dependent on O_2 concentration.
8. Plot of 1/rate versus $1/O_2$ at several different glucose concentrations give a series of parallel lines which indicates not only dependence on both O_2 and glucose concentration but also that the mechanism is a ping-pong bi-bi mechanism (Sec. 7.1.2). This may be indicated schematically as:

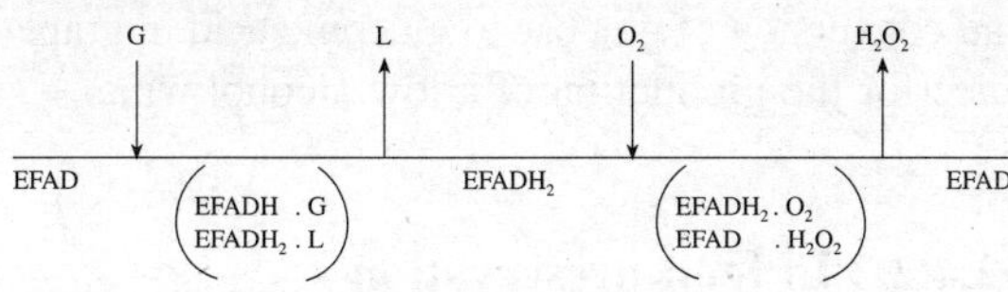

Where G and L stands for β-D-glucose and δ-D-gluconolactone, respectively.

On the basis of these observations, it was proposed that the entire mechanism of glucose oxidase-catalysed reaction consists of two half-reactions:

1. Reductive half-reaction in which E-FAD (oxidized form of the enzyme) is reduced to E-FADH$_2$ (reduced form of the enzyme) and the substrate (β-D-glucopyranose used) is oxidized to the product (δ-D-gluconolactone).

2. Oxidative half-reaction in which E-FADH$_2$ is oxidized to the reduced enzyme (E-FAD) and the electron acceptor O$_2$ is reduced to H$_2$O$_2$. δ-D-gluconolactone is hydrolysed non enzymatically to D-gluconic acid.

12.2 EFFECT OF pH

The observed effect of pH on activity of an enzyme may be influenced by the degree of saturation of enzyme with substrate. This is clearly the case of glucose oxidase.

To study the influence of substrate concentration on pH-activity profiles, a poor substrate, 2-deoxy-D-glucose was used. It was observed that:

a) With 2-deoxy-D-glucose, k_2 (conversion of Eox.S to Ered. P$_1$) is the rate determining step when the enzyme is saturated with both sugar and O$_2$ and the pH-activity is horizontal. Since the rate of conversion of Eox.S to Ered. P$_1$ is independent of pH, Eox.S cannot undergo any proton association-dissociation which affects the rate.

b) At saturating level of 2-deoxy-D-glucose and less than saturating levels of O$_2$, the rate of oxidation is dependent on O$_2$ concentration as controlled by k_4. The pH-activity profile shows that this step is pH dependent. The reason for being dependent on pH is the

ionization of the enzyme's prototropic group of pK 6.90. Combination with O$_2$ requires that the group with pK 6.90 be protonated in order, for the O$_2$, to combine with the enzyme and activity decreases above pH 6.0.

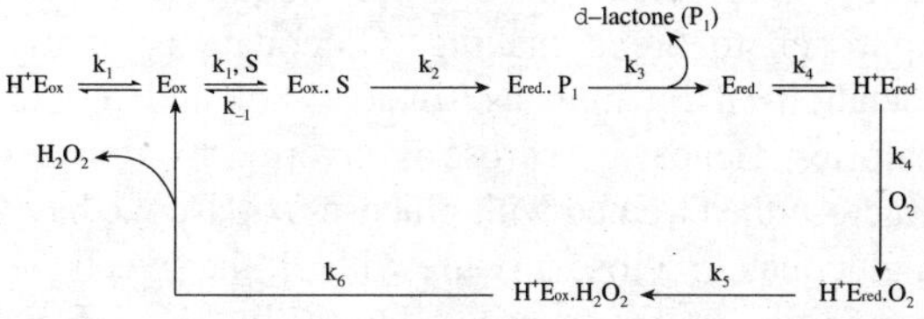

c) At less than saturating amounts of 2-deoxy-D-glucose and saturating amount of O$_2$, the pH-activity profile reflects the effect of pH on k_1 (pK = 5.0) and a sigmoidal curve is obtained.

d) When less than saturating amounts of both substrates are used a bell-shaped curve is obtained which reflects the pH dependence of both k_1 and k_4.

12.3 ACTIVITY ASSAY

Glucose oxidase activity can be followed by several methods. The two well-known methods are:

1. Since O$_2$ is one of the reactants its rate of disappearance from the reaction can be followed by the use of a manometric technique or with an O$_2$-sensitive electrode.

2. Activity of glucose oxidase can be determined by incorporation of peroxidase and chromogen, such as *o*-dianisidine, into the reaction mixture. Peroxidase utilizes the H$_2$O$_2$ produced to oxidize the chromogen to a coloured compound which can be determined spectrophotometrically.

12.4 APPLICATIONS IN FOOD PROCESSING

12.4.1 Production of Gluconate and δ-Gluconolactone

Glucose oxidase is used to manufacture gluconate and δ-gluconolactone which are now permitted food additives. Gluconate is an excellent chelator of calcium ions. Glucono δ-lactone in water produces gluconic acid so it can be used to initiate setting processes with alginate and high methoxyl pectin. It also has the stability to produce CO_2 with sodium bicarbonate, once mooted as a new approach to bread making. Gluconate is hardly metabolized by humans so does not add to our calories. Hence, if sucrose is inverted to glucose/fructose then treated with glucose oxidase we have a gluconate/fructose mixture with all the sweetness originally in the sucrose but only half of its calories.

12.4.2   Applications in Combination with Catalase

Glucose oxidase has been used in combination with catalase in some food industries. The presence of catalase in glucose oxidase preparations prevents the buildup of H_2O_2 from the oxidation of glucose.

12.4.2.1   In desugaring of glucose-rich food to prevent Maillard reaction

The desugaring of eggs prior to spray drying is one of the best known and earliest successful uses of glucose oxidase/catalase (GOX/CAT). In this application, GOX/CAT is used to oxidize the glucose present in eggs, thereby preventing Maillard browning in the finished dry product.

GOX/CAT can be used to desugar liquid whole eggs or the separated yolk and white (albumen). The pH of the albumen is adjusted to 6.8-7.0 to optimize the performance of the enzyme system. The pH adjustment is made with the slow addition of citric acid. After acidification H_2O_2 is added. Once the peroxide is blended well into the albumen, glucose oxidase is added. This dose is based on a 16 hr desugaring process. The desugaring process for egg yolk is similar to the procedure for albumen but there is no acid addition and the process is completed in about 4 hours.

GOX/CAT is also used for the glucose reduction in potatoes for minimization of browning

and conversion of glucose to gluconic acid in grape juice for the production of a low alcohol wine.

12.4.2.2 In food preservation

The glucose oxidase enzyme has been employed in food preservation since 1950s, mainly because of its desugaring and antioxidant capabilities.

12.4.2.2a Deoxygenation of beverages

GOX/CAT has been used for many years to deoxygenate beverages. One of the best markets for GOX/CAT for deoxygenation of beverages is the production of lemon or orange soft drinks sold in countries in which the product is exposed to light during distribution. In these situations, the use of the traditional oxygen scavengers and antioxidants, ascorbic acid, would bleach the beverages when exposed to light. To avoid a white citrus-flavoured soft drink, GOX/CAT is used instead of ascorbic to scavenge the oxygen in these products. Depending on the temperature, pH, and initial dissolved oxygen content, this dose can remove the dissolved oxygen within 3 hrs. The headspace oxygen in the sealed bottle or canned beverage is also scavenged over time, but its rate of removal depends on the rate of diffusion into the beverage.

12.4.2.2b In sea food preservation

The particular advantages in utilizing GOX/CAT for food preservation have been the antibacterial effects the enzyme produces, but more recently significant applications have been found in sea foods due to its antioxidant effects.

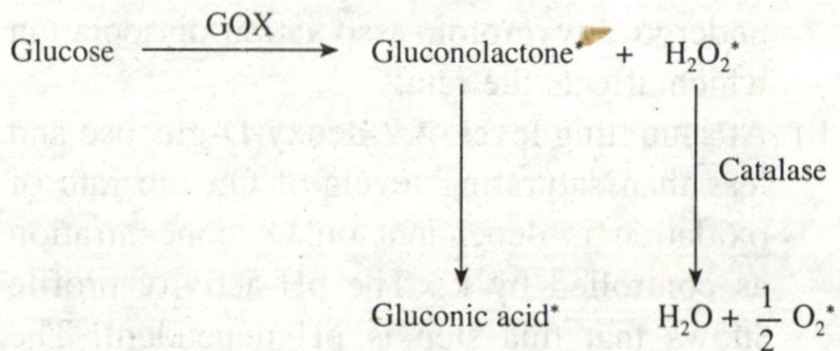

(Asterisks indicate potential antibacterial substances)

Since one of the proposed courses of preservative action by the GOX/CAT treatment has been thought to be oxygen depletion around the microenvironment of the fish, the enzyme system was studied as to its effect on the unsaturated fatty acids of fatty fish. The enzyme treatment maintained the integrity of these important fatty acids when compared to the control. Therefore, in addition to the antibacterial effect of the GOX/CAT system, these results demonstrated the potential of enzyme treatment to reduce lipid oxidation in fatty fish.

The potential of the GOX/CAT system in preserving shrimp has also been studied. This treatment not only extends shelf life, but prevents the serious discolouration known as black spot. This development of black spot in shrimp is due to the presence of an enzyme, polyphenol oxidase. The enzyme treatment depresses both autolytic and bacterial degradation. The main inhibitory effect of the enzyme system on polyphenol oxidase is via oxygen removal.

SUMMARY

Glucose oxidase (α-D-glucose: oxygen 1-oxidoreductase, E.C.1.1.3.4) catalyses the oxidation of α-D-glucose to δ-D-glocuno-1,5-lactone coupled with the reduction of O_2 to H_2O_2.

The enzyme is found in a number of fungal sources. The enzyme from different fungi share certain similarities in that all are dimers, composed of two identical subunits, have 2 moles of firmly bound FAD per mole of protein, and contain 11-20 per cent carbohydrates.

The observed effect of pH on activity of glucose oxidase is influenced by the degree of saturation of enzyme with substrate. Glucose oxidase activity can be followed either by measuring rate of disappearance of O_2 or by measuring rate of appearance of H_2O_2.

Glucose oxidase is used to manufacture gluconate and δ-gluconolactone which are now permitted food additives. Glucose oxidase has been used in combination with catalase in some food industries. The presence of catalase in glucose oxidase preparations prevents the buildup of H_2O_2 from the oxidation of glucose. The desugaring of eggs prior to spray drying is one of the best known and earliest successful uses of glucose oxidase/catalase GOX/CAT. The glucose oxidase enzyme has been employed in food preservation since 1950s, mainly because of its desugaring and antioxidant capabilities. GOX/CAT has been used for many years to deoxygenate beverages and as an antioxidant in sea foods.

RECOMMENDATIONS

1. Baldwin, R.R., Cambell, H.A., Theissen, R., and Lorant, G.J. (1953), 'The use of glucose oxidase in processing foods with special emphasis on desugaring of egg whites', *Food Technol.*, 7, pp. 275-82.
2. Barker, S.A. (1986), 'Extending enzyme applications in the food industry', *Chemical aspects of food enzymes* (A.T. Andrews edn.), pp. 137-48, University of Reading, Reading, England.
3. Bentley, R. (1963), 'Glucose Oxidase', *The Enzymes*, 2nd edn., 7, pp. 575-86.
4. Bright, H.J., and Appleby, M. (1969), 'The pH dependence of the individual steps in the glucose oxidase reaction', *J.Biol. Chem.*, 244, pp. 3625-34.
5. Bright, H.J. and Gibson, Q.H. (1967), 'The oxidation of 1-deuterated glucose by glucose oxidase', *J.Biol. Chem.*, 242, pp. 994-1003.
6. Gibson, Q.H., Swoboda, B.E.P., and Massey, V. (1964), 'Kinetics and mechanism of action of glucose oxidase', *J. Biol. Chem.*, 239, pp. 3927-34.
7. James, T.L., Edmondson, D.E., and Husain, M. (1981), 'Glucose oxidase contains a disubstituted phosphorus residue. Phosphorus-31 nuclear magnetic resonance studies of the flavin and nonflavin phosphate residues', *Biochemistry*, 20, pp. 617-21.
8. Kosikowski, F.V. (1977), 'Control of spoilage bacteria in cheese milk', *Cheese and fermented milk foods*, pp. 292-303, Kosikowski and Associates, Brooktondale, New York.

13 Catalase

Catalase (hydrogen peroxide: hydrogen-peroxide, E.C.1.11.1.6) is found in animals, plants and micro-organisms and has been purified from all three sources.

All catalases appear to have molecular weights of approximately 240,000 and to contain four protohemin groups per molecule. The molecule is composed of four subunits each of which contains a protohemin group. In protohemin, four of the six coordination bonds are taken up in interaction with the pyrrole ring nitrogens. The other two, X and Y, are occupied by water molecules or OH^- depending on the pH. In peroxidase, one of the two remaining coordination bonds (X and Y) appears to be a carboxyl group of the protein while the other is coordinated to an amino group or to a water molecule. The four active sites of catalase appear to function independently of each other. Since the prosthetic group of catalase is ferriprotoporphyrin III (Figure 13.1), this enzyme has an absorption maximum at 406 nm.

13.1 REACTION MECHANISM

The overall reaction involves the enzyme reacting with 2 moles of H_2O_2 to give water and oxygen as the final products.

$$2H_2O_2 \xrightarrow{\text{catalase}} 2H_2O + O_2$$

The chemistry of the reaction mechanism implies that H_2O_2 is involved in both oxidizing and reducing activities, with a two-electron transfer in each of the processes. This is shown in the following equations:

1) Oxidation of native ferric enzyme to catalase compound I, with the substrate H_2O_2 reduced to H_2O.

$$\text{En-Fe (III)} + H_2O_2 \longrightarrow \text{En-Fe (IV)} == O^+ + H_2O$$

2) Reduction of compound I back to the native ferric enzyme, with a second molecule of H_2O_2 oxidized to O_2.

$$\text{En-Fe (IV)} == O^+ + H_2O_2 \longrightarrow \text{En-Fe (III)} + O_2 + H_2O$$

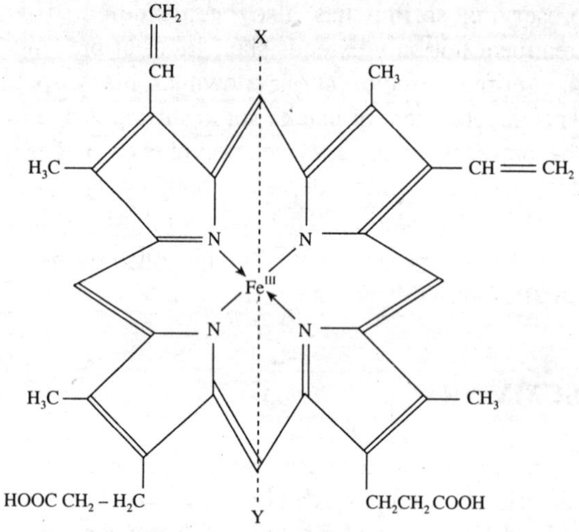

Fig. 13.1 *Structure of ferriprotoporphyrin III (protohemin).*

13.2 EFFECT OF pH AND TEMPERATURE

Catalase, perhaps as a result of its subunit structure, is not heat stable. It loses activity quite rapidly at 35°C. The catalytic activity is independent of pH over the range of 3 to 9. Above pH 9 and below pH 3 there is a decrease in activity of the enzyme. At alkaline pH the enzyme dissociates into subunits while at acidic pH the protohemin group dissociates from the protein.

13.3 ACTIVITY ASSAY

Catalase activity can be measured spectrophotometrically, manometrically and titrimetrically.

Hydrogen peroxide has a molar extinction coefficient of $52M^{-1}cm^{-1}$ at 235 nm and its disappearance can be followed by a continuous spectrophotometric method. Since O_2 is a product of the reaction, catalase activity can be determined readily by manometric techniques. The titrimetric method involves removal of aliquots of the reaction and titration of the remaining hydrogen peroxide with standard potassium permanganate solution in an acid solution.

13.4 APPLICATIONS IN FOOD PROCESSING

As mentioned earlier, catalases are used in the food industry for the removal of H_2O_2, the most well-known example being the oxidation of glucose by glucose oxidase and the coupled removal by catalases of the hydrogen peroxide. In addition to the applications discussed in chapter 12, following are the significant applications of catalases in food industries.

13.4.1 In Production of Cheese

Pasteurization is one of the methods used to ensure the microbiological safety of milk. When milk is being used for cheese production, this heat treatment adversely affects the flavour quality of the finished cheese. An alternative method permitted by the FDA is the use of up to 0.05 per cent of H_2O_2 for the chemical sterilization of milk for cheddar, washed curd, soaked curd, granular and stirred curd. Sufficient catalase to eliminate residual peroxide is required when using H_2O_2. The residual H_2O_2 is decomposed by catalase added to the treated milk.

The major benefit attributed to the use of catalase treatment of cheese milk is the improved flavour of the finished product. The flavours of the cheese made from catalase-treated cheese milk

are closer to that of cheese made with raw milk without the microbial problems associated with the use of raw milk.

13.4.2 In Whey as Bleaching Agent

H_2O_2 is used as a bleaching agent for coloured cheese whey. Catalase is used by the whey industry to dissipate residual H_2O_2. The procedure for removal of the residual peroxide with catalase is similar to that used for cheese milk.

SUMMARY

Catalase (hydrogen peroxide: hydrogen-peroxide, E.C.1.11.1.6) is found in animals, plants and micro-organisms and has been purified from all three sources.

All catalases appear to have molecular weights of approximately 240,000. The molecule is composed of four subunits each of which contains a protohemin group. The four active sites of catalase appear to function independently of each other. The catalytic reaction involves the enzyme reacting with 2 moles of H_2O_2 to give water and oxygen as the final products.

$$2H_2O_2 \xrightarrow{\text{catalase}} 2H_2O + O_2$$

Catalase is not heat stable. The catalytic activity is independent of pH over the range of 3 to 9.

Catalase activity can be followed by measuring the rate of disappearance of hydrogen peroxide. The titrimetric method involves titration of the remaining hydrogen peroxide with standard potassium permanganate solution in an acid solution. Since O_2 is a product of the reaction, catalase activity can be determined readily by manometric techniques.

Catalases are used in the food industry for the removal of H_2O_2, the most well-known example being the oxidation of glucose by glucose oxidase and the coupled removal by catalases of the hydrogen peroxide.

RECOMMENDATIONS

1. Dounce, A.L. (1983), 'A proposed mechanism for the catalase action of catalase', *J. Theor. Biol.*, 105, pp. 553-67.
2. Hillar, A., and Nicholls, P. (1992), 'A mechanism for NADH inhibition of catalase compound II formation', *FEBS Lett.*, 314, pp. 179-82.
3. Kosikowski, F.V. (1977), 'Control of spoilage bacteria in cheese milk', *Cheese and Fermented Milk Foods*, pp. 292-303, Kosikowski and Associates, Brooktondale, New York.
4. Miles Laboratories (1984), Catalase L, Application information, Miles Laboratories, Elkham, Indiana.

14 Peroxidase

Peroxidase is found widely distributed in higher plants (horseradish, turnip, fig sap, tobacco and potatoes) and micro-organisms.

14.1 CLASSIFICATION ON THE BASIS OF PROSTHETIC GROUP

On the basis of prosthetic group, the peroxidases may be classified as follows:

1. Iron-containing peroxidases

1a. Ferriprotoporphyrin peroxidases

This group includes peroxidases from higher plants, animals (iodine peroxidase of thyroid) and micro-organisms (cytochrome c peroxidase of yeast). These peroxidases contain ferri-protoporphyrin III as the prosthetic group which can be removed from the protein moiety on treatment with acidic acetone.

1b. Verdoperoxidases

These are found in milk (lactoperoxidase) myelocytes (myeloperoxidase) and in miscellaneous tissues.

2. Flavoprotein peroxidases

Flavoprotein peroxidases have been purified from several *Streptococci* including *Streptococcus faecalis* and from several animal tissues. The prosthetic group of these peroxidases is FAD. The remaining discussion will be focused on the well-studied horseradish peroxidase.

Horseradish peroxidase consists of many isoenzymes; a total of 42 has been reported. The isoenzyme C (pI 8.9) consists of a prosthetic group, (ferriprotoporphyrin), 2 calcium atoms, and 308 amino acid residues with 4 disulphide bonds and a calculated MW of 33,890 based on amino acid composition.

The native enzyme has the ferric ion coordinated to the four nitrogens of the pyrrole ring of the protoporphyrin. The fifth coordination is occupied by an imidazole ligand, the proximal His 170. The sixth coordination is vacant.

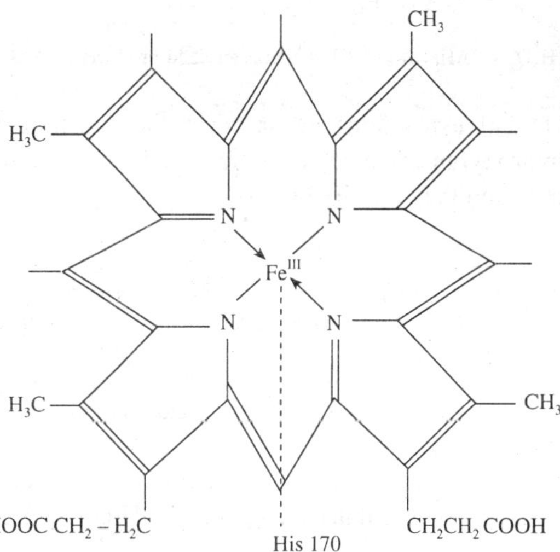

Fig. 14.1 *Horseradish peroxidase has ferric ion coordinated to the four nitrogens of the pyrrole ring and to the His 170.*

The enzyme contains 2 moles of calcium per mole of enzyme with one of the calcium atoms being essential for correct folding conformation. This calcium at the high-affinity binding site, is responsible for maintaining the protein structure around the haem group.

14.2 REACTION MECHANISM

Peroxidase catalyses a reaction in which hydrogen peroxide acts as the acceptor and another compound, AH_2, acts as the donor of hydrogen atoms. Molecular oxygen is not a product of the reaction.

$$H_2O_2 + AH_2 \longrightarrow 2H_2O + A$$

Peroxidase catalysis is associated with four types of activity : Peroxidative, oxidative, catalactic and hydroxylation.

Under the usual assay conditions where a phenolic substrate is used, only the peroxidative reaction is of importance. Peroxidative reactions occur when p-cresol, guaiacol, resorcinol, aniline, etc., are used as substrate.

$$H_2O_2 + 2AH_2 \longrightarrow 2H_2O + \text{polymerized products (HAAH)}$$

Oxidative reactions occur when the substrate is hydroxyfumaric acid, ascorbic acid, hydroquinone, etc., and they require O_2.

$$\begin{array}{c} \text{HO} - \text{C} - \text{COOH} \\ \| \\ \text{HOOC} - \text{C} - \text{OH} \end{array} + O_2 \longrightarrow \begin{array}{c} \text{O} = \text{C} - \text{COOH} \\ | \\ \text{O} = \text{C} - \text{COOH} \end{array} + 2H_2O_2$$

Dihydroxyfumaric acid   Diketosuccinic acid

In the absence of a hydrogen donor, peroxidase behaves as a catalase, converting H_2O_2 to H_2O and O_2

$$2H_2O_2 \longrightarrow 2H_2O + O_2$$

In the presence of certain hydrogen donors, particularly dihydroxyfumaric acid, and molecular oxygen, peroxidase can hydroxylate a variety of aromatic compounds including tyrosine, phenylalanine, p-cresol, benzoic and salicylic acid.

On the basis of the experimental data available the following general mechanism for the action of peroxidase is proposed. In the first step hydrogen peroxide replaces water at one of the available coordination bonds of the iron of the protohemin to form enzyme-substrate complex (per-Fe(III).H_2O_2). The second step involves the formation of compound I (Per-Fe(V)=O). In the third step, compound I reacts with an exogenous donor, if available, to give compound II (Per-Fe(IV)-OH) and a free radical from the hydrogen donor. In the fourth step, compound II reacts with a second molecule of the hydrogen donor to regenerate the original enzyme (Per-Fe(III).H_2O) and a free radical from the hydrogen donor.

Peroxidase is stable to 65-70°C at pH near 7.0 Peroxidase is also stable at alkaline conditions.

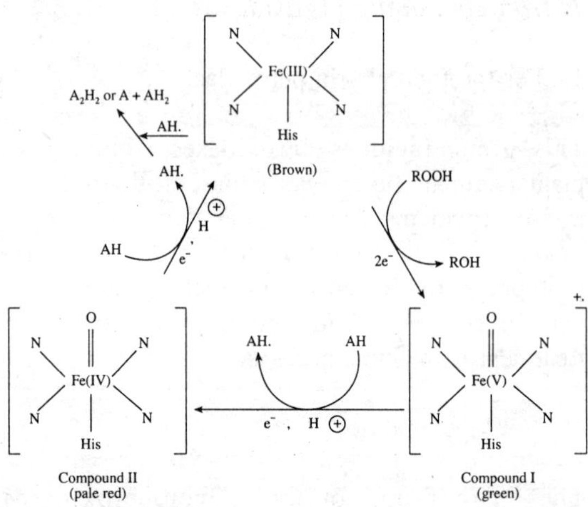

Fig. 14.2 *The general mechanism of horseradish peroxidase.*

14.3 ACTIVITY ASSAY

Assays of peroxidase activity involves the use of various hydrogen donors such as guaiacol, pyrogallol, uric acid and dehydroxyphenylalanine.

The guaiacol assay is very simple and readily followed in a continuous fashion in a reading spectrophotometer. Guaiacol and hydrogen peroxide are added to a buffer, usually at pH 7.0

and the reaction is initiated by addition of peroxidase. The increase in absorbance at 470 nm is measured as a function of time. The major product of the reaction is tetraguaiacol. From the molar extinction coefficient, ϵ, of tetraguaiacol at 470 nm and from the change in absorbance, the amount of hydrogen peroxide consumed in the reaction can be calculated.

14.4 APPLICATIONS IN FOOD

14.4.1 Peroxidase Activity and Food Quality

During the ripening of fruits, particularly during the climatic changes, it has been found that peroxidase activity increases along the activity of other enzymes, such as polygalacturonases and cellulases which are normally associated with the ripening process. Indeed, it has been suggested that peroxidase activity could be a possible parameter in the ripening of golden delicious apple. Also, after harvesting, the peroxidase activity of apricot has been reported to increase four to five fold.

Peroxidases promote a large number of reactions and therefore have versatility not surpassed by any other enzyme. Peroxidase does not exist as a single enzyme in fruits and vegetables, like many other plant enzymes, peroxidase activity is found in the form of a number of discrete isoenzymes, the isoperoxidases. There are many naturally occurring phenolic compounds in all plant tissues, each of which can be oxidized by peroxidase in the presence of small amounts of hydroperoxide. Due to the diversity of compounds which are susceptible to oxidations catalysed by peroxidases, the range of products formed is very extensive. Consequently, it is difficult to correlate perceived post-harvest changes, like loss of flavour and texture, with specific chemical reactions. Also, the plant tissues contain substrate-specific peroxidases, namely ascorbic acid peroxidase and glutathione peroxidase. Although the role of these enzymes is primarily the removal of hydrogen peroxide, the presence of ascorbic acid peroxidase in foods is likely to lead to additional loss of vitamin C.

The relationship of peroxidase activity to off-flavours and off-colours in raw and unblanched vegetables is still largely emperical. In post-harvest plant foods, peroxidases are believed to be responsible for deterioration in flavour, colour, texture and nutritional qualities of raw and processed foods. A relationship between peroxidase activity and off-flavour development in peas is well established. Active enzyme can spoil fruits and vegetables at temperatures as low as -18°C and at low moisture levels, the development of off-flavours is often associated with the oxidizing effect of peroxidases on indigenous lipids and the phenolic constituents of foods. For orange juice, it has been suggested that the significant negative correlation between peroxidase activity and the flavour scores of high- and low-yield orange juice indicates that it might be possible to use peroxidase activity as an index of potential adverse flavour. For beans, the qualitative peroxidase test was found to be the best index of adequate blanching.

14.4.1.1 Oxidation of indole-3-acetic acid

Peroxidases are believed to play an important role in fruit ripening. It is generally accepted that the degradation of indole-3-acetic acid (IAA) by peroxidases controls the *in vivo* concentration of this phytohormone in plants, although other enzymes may also catalyse the degradation of IAA. Inactivation of the plant growth regulator occurs mainly by an oxygen-consuming oxidation catalysed by peroxidases.

14.4.1.2 Oxidation of chlorophyll

Peroxidases may play important roles in the colour changes associated with the ripening of fruits and vegetables. Although two other enzymes, namely chlorophyllase and chlorophyll oxidase, are known to catalyse the degradation of chlorophyll. There is increasing circumstantial evidence that peroxidase plays a role in leaf senescence and the associated yellowing of fruits. For instance, it has been claimed that peroxidase activity and levels of hydrogen peroxide increase during the senescence and ripening of fruits.

14.4.1.3 Lignin biosynthesis

Lignin is an abundant heterogenous polymer that forms a three-dimensional network and is generally regarded as being responsible for toughness in vegetables and stringiness in beans. The two reactions involved in the lignification process are oxidative polymerization of coniferyl alcohol and the generation of hydrogen peroxide at the expense of NADH oxidation. Peroxidase first catalyses the oxidation of coniferyl alcohol to form phenoxy radicals which then polymerize non-enzymically to form lignin, a strong, resistant polymer with a high proportion of ether bonds and methoxy groups. It is claimed that lignification is always accompanied by an increased activity of most isoperoxidases.

Strengthening of cell walls by lignification during the growth and maturation of plants can largely account for the increased hardness and toughness in green beans. During the maturation of seeds, peroxidase is instrumental in the conversion of soluble phenolics to insoluble lignin polymers.

Many tropical fruits are susceptible to chill injury and peroxidase may be involved in these commercially undesirable changes.

14.4.2 Industrial Application

Comprehensive knowledge of the occurrence and enzymic properties of isoperoxidases is a prerequisite to any improvement that may be achieved through genetic selection of cultivars for improved storage and processing. Further in the future, it will be possible by genetic engineering to modify the action and the amounts of peroxidases in order to control their action in post-harvest foods. One example would be the control of lignification by suppressing either the enzymic activity or biosynthesis of lignin peroxidase. It is perhaps conceivable that the extent of ripening may also be influenced by the control of the oxidative action of peroxidases on indoleacetic acid when the specific peroxidase for this phytohormone has been characterized. Likewise, it may also be possible to identify the isoperoxidases responsible for the peroxidatic oxidation of flavonoids, other phenolics and susceptible flavour substances and thus decrease the deleterious action of the enzymes.

Just based on our knowledge of the action of peroxidases on phenolics, dihydroxy compounds, like dihydroxyfumarate and heterocycles like indoleacetic acid, a likely use of peroxidases is to provide sources of both simple and complex oxidized aromatic intermediates in the form of hydroperoxides or hydroxy compounds. The degradation of chlorophyll, catalysed by peroxidases, also points towards a possible use for the degreening of, as yet, unattractive leaf protein and other proteins of photosynthetic origin. Some more unusual peroxidases are already known that catalyse the oxidation of halogens to form hypochlorous and hypobromous acids and therefore, there is a known potential for their use as oxidizing agents. Prior oxidation of aromatics in waste effluents by peroxidases may also be very beneficial in reducing the oxygen demands that such products place on lakes and rivers.

Lignin peroxidases are likely to have a major part to play, not only for the possible conversion of degraded lignins to more valuable substances and as providers of chemical intermediates, but also in the utilization of agricultural products such as waste plant residues from sugar cane, sugar beet and other leafy crops.

SUMMARY

Peroxidase is found widely distributed in higher plants (horseradish, turnip, fig sap, tobacco and potatoes) and micro-organisms.

On the basis of prosthetic group, the peroxidases may be classified into (1) Iron-containing peroxidases and (2) Flavoprotein peroxidases. Former group of peroxidases contain ferri-protoporphyrin III as the prosthetic group in which ferric ion is coordinated to the four nitrogens of the pyrrole ring. The prosthetic group of flavoprotein peroxidases is FAD.

Peroxidase catalyses a reaction in which hydrogen peroxide acts as the acceptor and another compound, AH_2, acts as the donor of hydrogen atoms.

$$H_2O_2 + AH_2 \longrightarrow 2H_2O + A$$

Peroxidase catalysis is associated with four types of activity: Peroxidative, oxidative, catalactic and hydroxylation. Peroxidative reactions occur when p-cresol, guaiacol, resorcinol, aniline, etc., are used as substrate. Oxidative reactions occur when the substrate is hydroxyfumaric acid, ascorbic acid, hydroquinone, etc., and they require O_2. Catalytic reactions occur in the absence of a hydrogen donor, converting H_2O_2 to H_2O and O_2. Hydroxylation reaction occurs in the presence of certain hydrogen donors, particularly dihydroxyfumaric acid and molecular oxygen. In the presence of these hydrogen donors peroxidase can hydroxylate a variety of aromatic compounds including tyrosine, phenylalanine, p-cresol, benzoic and salicylic acid.

Assay of peroxidase activity involves the use of various hydrogen donors such as guaiacol. The major product of the reaction is tetraguaiacol. The increase in absorbance at 470 nm is measured as a function of time.

Peroxidases are believed to play an important role in fruit ripening.

RECOMMENDATIONS

1. Adediran, S.A., and Dunford, H.B. (1983), 'Structure of horseradish peroxidase compound I, Kinetic evidence for the incorporation of one oxygen atom from the oxidizing substrate into the enzyme', *Eur. J. Biochem.*, 132, pp. 147-50.
2. Bedford, C.L. and Joslyn, M.A. (1939), 'Enzyme activity in frozen vegetables: String beans', *Ind. Eng. Chem.*, 31, pp. 751-8.
3. Brennan, T. and Frenkel, C. (1977), 'Involvement of hydrogen peroxide in the regulation of senescence in pear', *Plant Physiol.*, 59, pp. 411-416.
4. Bruemmer, J.H., Roe, B. and Bowen, E.R. (1976), 'Peroxidase reactions and orange juice quality', *J. Food Sci.*, 41, pp. 186-9.
5. Gorin, N. and Heidema, F.T. (1976), 'Peroxidase activity in golden delicious apples as a possible parameter of ripening and senescence', *J. Agric. Food Chem.*, 24, pp. 200-201.
6. Haschke, R.H., and Friedhoff, J.M. (1978), 'Calcium-related properties of horseradish peroxidase', *Biochem. Biophys. Res. Comm.*, 80, pp. 1039-42.
7. Joslyn, M.A. (1949), 'Enzymatic activity in frozen vegetable tissue', *Adv. Enzymol.*, 9, pp. 613-52.
8. Lauriere, C. (1983), 'Enzymes and leaf senescence', *Physiol. Veg.*, 21, pp. 1159-77.
9. Matile, P., Ginsburg, S., Schellenberg, M. and Thomas, H. (1987), 'Catabolites of chlorophyll in senescent leaves', *J. Plant Physiol.*, 129, pp. 219-28.
10. Ogawa, S., Shiro, Y., and Morishima, I. (1979), 'Calcium binding by horseradish peroxidase C and the heme environmental structure', *Biochem. Biophys. Res. Comm.*, 90, pp. 674-78.
11. Reinecke, D.M. and Bandurski, R.S. (1988), 'Oxidation of indole-3-acetic acid to oxindole-3-acetic acid by an enzyme preparation from Zea mays', *Plant Physiol.*, 86, pp. 868-72.

15 Polyphenol Oxidase

Polyphenol oxidase (1,2-benzenediol: oxygen oxidoreductase; EC. 1.10.3.1) is frequently called tyrosinase, polyphenolase, phenolase, catechol oxidase, cresolase, or catecholase. The tyrosinase came from tyrosine being used first as a substrate.

The enzyme is present in all plants, but it is in particularly high concentration in mushrooms, potato tubers, peaches, apples, bananas, avocados, tea leaves, coffee beans and tobacco leaves.

15.1 HETEROGENEITY AND MOLECULAR STRUCTURE

Gel electrophoresis, isoelectric focusing, immunological analysis and gel filtration of many PPO preparations indicate that the enzymes exists in multiple forms. Enzyme multiplicity has been attributed to association-dissociation phenomena of similar and/or different subunits, polymerization with phenolics, glycosylation, and proteolytic action during isolation/purification of the enzyme, as well as interaction between enzyme and nonenzymatic proteins. These forms may also be interconvertible (depending on ionic strength, pH, enzyme concentration and presence of dissociating agents, etc.) and differ in their substrate specificity, pH optima, thermal stability and response to inhibitors or chemical reagents.

Reported molecular weights of multiple forms of PPO, as determined by SDS electrophoresis or gel filtration, extend over a wide range, 12,000-400,000. Often the pattern of multiple forms was found to vary within the same species or even cultivars depending on the time of harvest, storage conditions, etc. Thus, while Harel and Mayer observed three forms of catechol oxidase in apple fruit (30-40, 60-70 and 120-130 K), Janowitz-Klapp in 1989 found a single PPO component (mol. wt 46,000) for another apple cultivar. Multiplicity of molecular weights has also been observed for PPO from avocado (14, 28, 56, 112, and 400 K), potato tubers (aggregates of 2, 4, 8, and 16 of a 36-K subunit) and sugarcane (32 and 130 K), as assessed by gel filtration chromatography; wild rice (35, 42, 48, and 116 K); kiwi fruit (15, 20, 25 and 45 K) and wheat (23.5 and 30 K) as revealed by SDS-PAGE. These results seem to indicate that the molecular weight of the basic subunit of PPO from plants and other sources ranges between 30 and 45 K.

Published amino acid analyses of phenolases from several plant sources show that there are similarities between the enzymes as shown by their content of hydrophobic and sulphur-containing amino acids. Moreover, all plant PPOs exhibit similarities in their content of basic amino acids (Arg, His, Leu).

Polyphenol oxidases are copper-containing enzymes. With respect to the prosthetic group, the current consensus is that the functional unit of the *Neurospora* and mushroom enzymes contain a pair of copper ions per single polypeptide chain, each of which is liganded to three histidine residues (Figure 15.1).

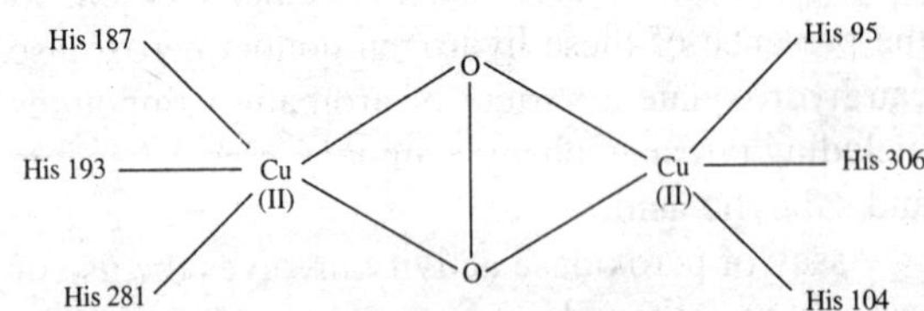

Fig. 15.1 *Coordination of copper to six histidine residues in the active site of Neurospora polyphenol oxidase.*

Therefore, any mechanism of action for polyphenol oxidases must include the role of the coppers. This appears to fit the data for most polyphenol oxidases.

15.2 REACTIONS CATALYSED

Unlike most enzymes, polyphenol oxidase can catalyse two quite different types of reactions, both of which involve phenolic compounds. These reactions involve hydroxylation of monophenols to give o-diphenols and the removal of hydrogens from o-diphenol to give o-quinone.

15.2.1 Hydroxylation

15.2.1.1 Mechanism of reaction

Hydroxylation of monophenols is frequently referred to as cresolase activity since p-cresol is often used as substrate.

$$p\text{-cresol} + O_2 + BH_2 \longrightarrow 4\text{-methylcatechol} + B + H_2O \quad (1)$$

p-cresol   4-methylcatechol

In the equation above, BH_2 is an o-diphenol required as a cofactor. The polyphenol oxidases that perform monohydroxylation reaction show hysteresis (lag phase), if BH_2 is not added initially to the reaction. The enzymes can slowly produce BH_2 by o-hydroxylation of the monophenol (equation 2).

$$ + O_2 + \longrightarrow + + H_2O \quad (2)$$

In the absence of added o-diphenol, there is a small amount of o-diphenol via equation 3.

$$ + E\text{-}2Cu^{\oplus} + O_2 + 2H^{\oplus} \longrightarrow + E\text{-}2Cu^{2\oplus} + H_2O \quad (3)$$

The o-diphenol functions to reduce the enzyme form $E\text{-}2Cu^{2+}$ to $E\text{-}Cu^{+}$ (equation 4) which can then hydroxylate the monophenol (equation 3).

$$ + E\text{-}2Cu^{2\oplus} \longrightarrow + E\text{-}2Cu^{\oplus} + 2H^{\oplus} \quad (4)$$

In the absence of added catechol, the overall reaction is:

$$ + O_2 \longrightarrow + H_2O \quad (5)$$

15.2.1.2 Assay of hydroxylation reaction

The rate of hydroxylation may be followed by measuring O_2 uptake, either in a respirometer or by use of an O_2-sensitive electrode. Both procedures are quite convenient and give a measure of activity of the enzyme; however, they also measure the O_2 uptake due to oxidation of the o-diphenol. A more specific method is to follow the reaction spectrophotometrically at a wavelength where only formation of o-diphenol is observed. Only initial rates are used to eliminate interference from later polymerization reactions which occur.

15.2.2 Oxidation of o-Diphenols (Dehydrogenation)

15.2.2.1 Mechanism of reaction

The second type of reaction catalysed by polyphenol oxidase is the oxidation of an o-diphenol to a benzoquinone (equation 6).

$$2 + O_2 \longrightarrow + 2H_2O \quad (6)$$

Catechol   *o*-Benzoquinone

The substrate most frequently used in the assay of the activity of polyphenoloxidases on *o*-diphenols is catechol. Thus, this activity is referred to as catecholase activity.

All polyphenol oxidases have activity on *o*-diphenols. Polyphenol oxidases from banana, tea leaf and tobacco leaf have been reported to have exclusive activity on *o*-diphenols and no ability to hydroxylate monophenols. Polyphenol oxidase from potato, apple, sugar beet leaf and broad bean leaf have both types of activity.

Oxygen is bound first to the two Cu(I) groups of deoxypolyphenol oxidase, to give oxypolyphenol oxidase in which the O_2 has the characteristics of a peroxide. The two Cu(II) groups of oxypolyphenol oxidase then bind the two hydroxyl groups of catechol, replacing the two water molecules or OH groups, to form the O_2-catechol- enzyme complex. The catechol is oxidized to benzoquinone, leaving the enzyme as *m*-polyphenol oxidase. Another molecule of catechol binds to *m*-polyphenol oxidase, and is oxidized to benzoquinone, in the process reducing the enzyme to deoxypolyphenol oxidase, thereby completing the cycle. Two catechol molecules are oxidized in a complete cycle, consuming two atoms of O_2 (Fig. 15.2).

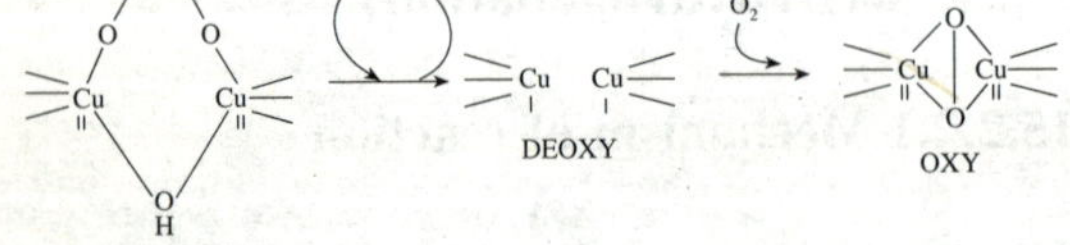

Fig. 15.2 *Proposed kinetic scheme depicting the mechanisms of o-diphenols (catechol).*

15.2.2.2 Assay of activity on *o*-diphenols

Several methods are available for following the activity of polyphenol oxidase on *o*-diphenols such as catechol. These include manometric, polarographic and spectrophotometric methods. The manometric and polarographic methods measure the oxygen consumption of the system. In spectrophotometric method the decrease in concentration of catechol is measured at 395 nm with respect to time.

15.3 pH OPTIMUM AND THERMAL STABILITY

The pH optimum of PPO from many plants sources, including apple, pear, peach, grape, banana, avocado, potato and mushroom, is generally between 5.0 and 7.0. It is also noteworthy that two pH optima have been found for some PPO preparations, with the activity being greater at the higher pH, e.g. grapes, 5.0 and 7.3; wheat, 5.3 and 5.9; apples, 5.2 and 7.3. PPO is often reported to be inactive at pHs below 4.0, thus providing a method for controlling browning reactions by lowering the pH. Other features which recur from various reports in the literature are that the pH-activity profiles vary with the cultivar and stage of maturity, enzyme purity, nature of the phenolic substrate and isoenzyme form.

Although dependent on the source, PPO is generally considered as an enzyme of low thermostability. Heat stability differs among cultivars and multiple forms of PPO from the same source as well as, between fruit tissue homogenates and their respective juices. Thermal treatments of relatively short duration at 70-90°C are sufficient to substantially reduce or completely eliminate the PPO activity in products of plant origin. In contrast, the enzyme is relatively stable under storage at subzero temperatures.

15.4 APPLICATIONS IN FOOD

15.4.1 PPO Activity and Food Quality

PPO is a very important enzyme in determining the quality and economics of fruit and vegetable storage and processing. Up to one-half of some fruits are lost because of browning. Bruises, cuts and other mechanical damages, that allow O_2 penetration, lead to rapid browning in many fruits and vegetables because of melanin formation. The colour of damaged products, the off-taste and the loss of nutritional quality are unacceptable to the consumer. Peaches, apricots, apples, grapes, bananas, strawberries and several tropical fruits and juices brown, as do the vegetables like potatoes. Black spot development in shrimps is also a major economic problem.

15.4.2 Applications in Food Processing

The action of PPO is desirable in tea, coffee, cocoa and black raisins. The melanins produced by PPO have potential as food colourants and act as excellent sun blockers when applied to the skin.

15.4.2.1 Tea manufacture

PPO plays a key role during fermentation of black tea by oxidizing polyphenols. The major group of polyphenols in tea are the catechins, which include, catechin, epicatechin, gallocatechin, epigallo-catechin, epicatechin gallate and epigallocatechin gallate. During fermentation the catechins are oxidized to the corresponding *o*-quinones, which then undergo secondary oxidation to form theaflavin and theaflavin gallate. The latter compounds are both yellow pigments which undergo oxidation to thearubigens, dark brown pigments responsible for the colour of black tea.

The quality of tea, based on sensory evaluation of colour and bitterness, has been correlated with total theaflavin content. Theaflavins, thearubigens, and caffeine are all essential ingredients in high-quality teas. PPO activity is directly related to high level of caffeine, crude fibre and theaflavins.

15.4.2.2 Cocoa processing

In chocolate production, PPO is one component responsible for the formation of flavour precursors, beginning in the oxidative phase of the fermentation and continuing into the drying phase.

Among the changes affecting flavour are reduction in the bitterness and astrigency that result from polymerized polyphenol-protein interactions. An increase in oxygen penetration into the cocoa bean mass during drying induces maximum oxidation of epicatechin and procyanidins, causing the production of melanin and melanoproteins that are responsible for the colour in brown chocolate.

SUMMARY

Polyphenol oxidase (1,2-benzenediol: oxygen oxidoreductase; EC. 1.10.3.1) is present in all plants, but it is particularly in high concentration in mushrooms, potato tubers, peaches, apples, bananas, avocados, tea leaves, coffee beans and tobacco leaves. The enzymes exist in multiple forms. As a prosthetic group, the enzyme contains a pair of copper ions per single polypeptide chain, each of which is liganded to three histidine residues.

Polyphenol oxidase can catalyse two quite different types of reactions, both of which involve phenolic compounds. These reactions involve hydroxylation of monophenols to give *o*-diphenols and the removal of hydrogens from *o*-diphenol to give *o*-quinone. The rate of hydroxylation may be followed by measuring O_2 uptake either in a respirometer or by use of an O_2-sensitive electrode. The activity of polyphenol oxidase on *o*-diphenols may be followed either by measuring the oxygen consumption of the system or by measuring the decrease in concentration of catechol at 395 nm with respect to time.

The pH optimum of PPO from many plants sources is generally between 5.0 and 7.0. PPO is generally considered as an enzyme of low thermostability.

PPO is a very important enzyme in determining the quality and economics of fruit and vegetable storage and processing. The action of PPO is desirable in tea, coffee, cocoa and black raisins. The melanins produced by PPO have potential as food colourants and act as excellent sun blockers when applied to the skin.

RECOMMENDATIONS

1. Lerch, K., and Ettlinger, L. (1972), 'Purification and characterization of a tyrosinase from *Streptomyces glaucescens*', *Eur, J. Biochem.*, 31, pp. 427-37.

2. Mayer, A.M. and Harel, E. (1979), 'Review: Polyphenol oxidases in plants', *Phytochemistry*, 18, pp. 193-215.

3. Robb, D.A. (1984), Tyrosinase, In copper proteins and copper enzymes, pp. 207-41, R. Lontie. CRC Press, Boca Raton, Florida.

4. Schoenbein, C.F. (1956), 'On ozone and oronic actions in mushroooms', *Phil.Mag.*, 11, pp. 137-141.

5. Sherman, T.O., Vaughn, K.C., and Duke, S.O. (1991), 'A limited survey of the phylogenetic distribution of polyphenol oxidase', *Phytochemistry*, 30, pp. 2499-506.

6. Solomon, E.I., Baldwin, M.J., and Lowrey, M.D. (1992), 'Electronic structures of active sites in copper proteins: Contributions to reactivity', *Chem. Rev.*, 92, pp. 521-42.

7. Sugumaran, M. (1988), 'Molecular mechanisms for cuticular sclerotization', *Adv. Insect Physiol.*, 21(9), pp. 179-231.

8. Vamos-Vigyazo, L. (1981), 'Polyphenol oxidase and peroxidase in fruits and vegetables', *CRC Critic. Rev. Food Sci. Nutr.*, 15, pp. 49-127.

9. Witkop, C.J., Jr. (1979), 'Depigmentations of the general and oral tissues and their genetic foundations', *Ala. J. Med. Sci.*, 16, pp. 331-43.

16 Lipoxygenase

Lipoxygenase (linoleate: oxygen oxidoreductase, EC. 1.3.11.12) is a dioxygenase that catalyses the oxygenation of polyunsaturated fatty acids (LH) containing a *cis,cis*-1,4-pentadiene system to hydroperoxides (LOOH). Lipoxygenase is found in a variety of plants particularly the legumes including peas, beans and peanuts, radishes and potatoes. The enzyme exists in multiple forms, three in soybeans and peas, and two in corn. The highest activity is found in soybeans.

Soybean lipoxygenase 1 (LOX-1) is composed of a single polypeptide chain of 838 amino acids, with MW 94,038, as deduced from the nucleotide sequence. The enzyme contains four Cys but no disulphides. LOX-2 has a MW of 97,036 with 865 amino acids, and contains 7 Cys and 15 Trp. The deduced amino acid sequence of LOX-3 consists of 857 amino acids with a MW of 96,663. There are 5 Cys and 14 Trp, but no disulphides in the enzyme molecule.

16.1 MECHANISM OF CATALYSIS

For the lipoxygenase-catalysed dioxygenation reaction, two mechanisms have been proposed: (1) formation of a pentadienyl radical and peroxy radical intermediate involving enzyme complexes of Fe(II) and Fe(III) species (the radical mechanism), (2) ferric ion-assisted deprotonation of the pentadiene and binding to the carbanion followed by insertion of dioxygen into the iron-carbon bond (the organoiron-mediated pathway).

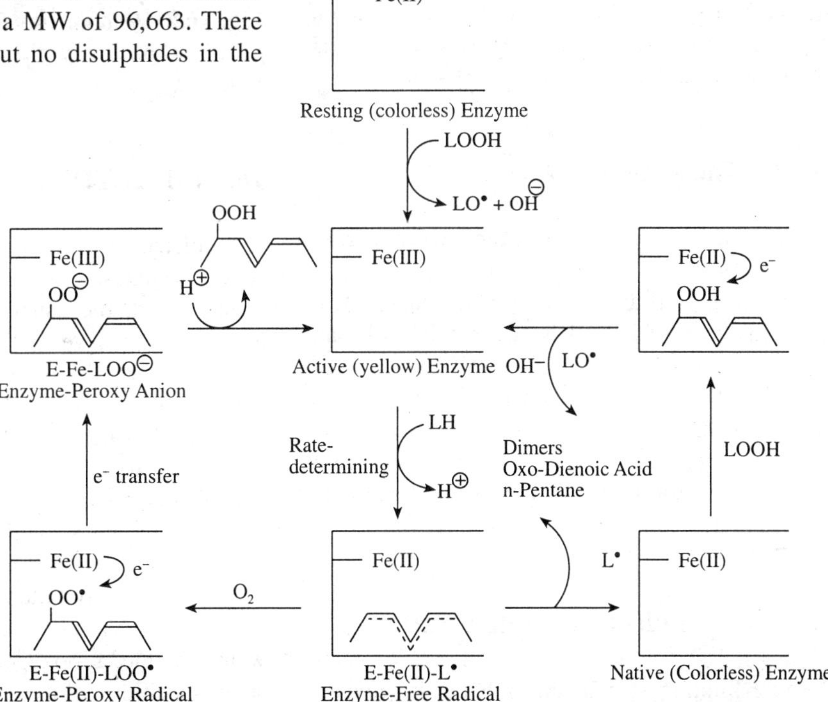

Fig. 16.1 *Reaction scheme for the activation of soybean lipoxygenase-1 and for the catalytic mechanism under aerobic and anaerobic conditions.*

16.1.1 Radical Mechanism

The reaction scheme outlined in figure 16.1 describes the radical reaction mechanism under both aerobic and anaerobic conditions for soybean LOX-1, the substrates are linoleic acid and oxygen.

16.1.1.1 The aerobic cycle

In the aerobic cycle, the Fe(III) in the active enzyme [E-Fe (III)] oxidizes the 1,4-diene of the substrate via a one-electron transfer that leads to the stereospecific H abstraction and subsequent formation of a pentadienyl radical [E-Fe(I)-L]. The abstraction of C11-H from linoleic acid by soybean LOX-1 has been shown to be the rate-limiting step in the overall reaction.

Oxygen then combines stereospecifically with the enzyme free-radical to form the enzyme-peroxy radical [E-Fe (II)-LOO$^\bullet$]. This oxygenation step is followed by a one-electron transfer in the reoxidation of the Fe(II), and formation of the enzyme-peroxyanion [E-Fe(II)-LOO$^-$]. Protonation of the enzyme-peroxy anion releases the LOOH and regenerates the active enzyme [E-Fe(III)].

16.1.1.2 The anaerobic cycle

Under anaerobic conditions, the enzyme-free-radical complex dissociates into the ferrous enzyme [E-Fe(II)] and the fatty acid radical [L$^\bullet$]. The breakdown of LOOH is also catalysed by an initial one-electron transfer from Fe(II). The alkoxy radical [LO$^\bullet$] formed can undergo a number of reactions, including reaction with more LOOH to form the peroxy radical [LOO$^\bullet$]. Rearrangement and decomposition of these fatty acid radicals yields oxo-dienoic acid and pentane.

16.1.2 Organoiron-mediated Pathway

The rate limiting step in this scheme involves a general base-catalysed abstraction of the proton at the methylene carbon, facilitated by a concomitant electrophilic attack of Fe (III) to the carbon. The resulting organoiron intermediate is carbanion in character, which reacts with dioxygen by σ-bond insertion assisted by an electron transfer from the olefin to the iron. The LOOH is formed by protonation of the peroxy anion (Figure 16.2).

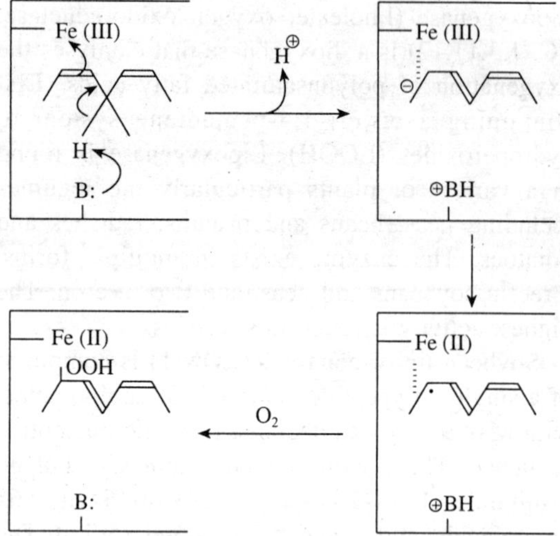

Fig. 16.2 *Organoiron-mediated oxygenation as a possible pathway for lipoxygenase-catalysed reaction.*

16.2 pH-ACTIVITY PROFILE

It is difficult to interpret the pH-activity profiles of lipoxygenase, partially due to insolubility of linoleic acid, the usual substrate, below pH 7.0. Most workers have found bell-shaped curves with maxima near 7.0-8.0 but there is a possibility that the decrease below pH 7.0 is due to insolubility of substrate. Addition of organic solvents or detergents to dissolve the substrate at low pH values often results in peculiar pH-activity profiles.

To overcome difficulties due to insolubility of substrate and presence of detergents, it has been proposed that octadeca-9, 12-dienyl sulphate (linoleyl sulphate) be used as a substrate since it is water soluble. Lipoxygenase appears to have the same specific activity on it as on linoleate. With this compound the pH-activity profile is a symmetrical bell-shaped curve with a pH maximum of 8.15.

16.3 APPLICATIONS IN FOOD

16.3.1 Lipoxygenase Activity and Food Quality

Lipid oxidation is of major importance for the quality of our foods and for our health. Lipid oxidation changes the aroma, flavour, taste, texture, colour and nutritional quality of foods. Peroxidation of lipids as a result of insufficient amounts of vitamin E, leads to major damage to tissues, especially to membranes.

Changes in aroma, flavour and taste of some foods frequently arise from products formed directly from oxidation of polyunsaturated lipids. Of even more concern are the free radical intermediates formed that can interact with and alter other constituents, such as proteins, amino acids, nucleic acids, carbohydrates, vitamins and pigments. Some of the products formed may be toxic, such as the potential carcinogen, malonaldehyde. Malonaldehyde may act as catalyst in formation of N-nitrosoamines in foods containing secondary amines and nitrite. Cholesterol oxidation product may be angiotoxic or even carcinogenic. Fatty acid hydroperoxides may be catalysts in the formation of N-nitrosopyrrolidine, for example in grilled or fried bacon.

Lipoxygenase is probably responsible for the degradation of chlorophyll in underblanched green beans, green peas and other leguminous foods.

16.3.2 Applications in Food Processing

One of the first uses of lipoxygenases in food processing was the bleaching of carotenoids in wheat dough. This is generally done by addition of small quantity of soybean flour, high in lipoxygenase activity.

Lipoxygenase, in the form of soybean flour, is also added to dough to improve the rheological properties, which results in larger loaf volumes and improved texture. Lipoxygenase also increases the ability of the dough to withstand overmixing. There are several hypotheses to explain this role. The most accepted hypothesis is that oxidation of the thiol groups of gluten by the lipid hydroperoxide formed is responsible for the improvement. This hypothesis is re-enforced in that lipoxygenase can replace the need for use of the oxidizing reagents, bromates and iodates. Some workers believe that lipoxygenase assist in the release of bound lipids. Lipoxygenase also appears to have a marked effect in retarding bread staling.

Lipoxygenase are responsible for some of the desirable flavours and aromas of vegetables and fruits. In tomatoes, the formation of 2-hexenal from polyunsaturated fatty acids is important. In fruits, the initial products of lipid oxidation by lipoxygenase are converted by other enzymes to alcohols and acids, which then form esters.

SUMMARY

Lipoxygenase (linoleate: oxygen oxidoreductase, EC 1.3.11.12) is a dioxygenase that catalyses the oxygenation of polyunsaturated fatty acids (LH) containing a *cis,cis*-1,4-pentadiene system to hydroperoxides (LOOH). Lipoxygenase is found in a variety of plants. The enzyme exists in multiple forms. The highest activity is found in soybeans.

For the lipoxygenase-catalysed dioxygenation reaction, two mechanisms have been proposed: (1) formation of a pentadienyl radical and peroxy radical intermediate involving enzyme complexes of Fe(II) and Fe(III) species (the radical mechanism), (2) ferric ion-assisted deprotonation of the pentadiene and binding to the carbanion followed by insertion of dioxygen into the iron-carbon bond (the organoiron-mediated pathway).

It is difficult to interpret the pH-activity profiles of lipoxygenase, partially due to insolubility of linoleic acid, the usual substrate, below pH 7.0.

Lipoxygenase is responsible for changes in aroma, flavour and taste of some foods that frequently arise from products formed directly from oxidation of polyunsaturated lipids. One of the first uses of lipoxygenases in food processing was the bleaching of carotenoids in wheat dough. Lipoxygenase, in the form of soybean flour, is also added to dough to improve the rheological properties. Lipoxygenase also appears to have a marked effect in retarding bread staling.

RECOMMENDATIONS

1. Chan, J.T. and Black, H.S. (1976), 'Distribution of cholesterol -5a, 6a-epoxide formation and its metabolism in mouse skin', *J. Invest. Dermat.*, 66, pp. 112-16.
2. Coleman, M.H. (1978), 'A model system for the formation of N-nitrosopyrrolidine in grilled or fried bacon', *J. Food Technol.*, 13, pp. 55-69.
3. Corey, E.J., and Nagata, R. (1987), 'Organoiron-mediated oxygenation of allylic organotin compounds. A possible model for enzymatic lipoxygenation', *J. Am. Chem. Soc.*, 109, pp. 8107-08.
4. Garssen, G.J., Vliegenthart, J.F.G., and Boldingh, J. (1971), 'An anareobic reaction between lipoxygenase, linoleic acid and its hydroperoxides', *Biochem. J.*, 122, pp. 327-32.
5. Lee, F.A. and Wagenknecht, A.C. (1958), 'Enzyme action and off-flavour in frozen peas. II. The use of enzymes prepared from garden peas', *Food Research*, 23, pp. 584-90.
6. Shibata, D., Steczko, J., Dixon, J.E., Andrews, P.C., Hermodson, M., and Axelrod, B. (1988), 'Primary structure of soybean lipoxygenase L-2', *J. Biol. Chem.*, 262, pp. 10080-85.
7. Smith, D.E., Buren, J.P. Van and Andrews, J.S. (1957), 'Some effects of oxygen and fat upon the physical and chemical properties of flour doughs', *Cereal Chem.*, 34, pp. 337-49.
8. Sullivan, B., Near, C. and Foley, G.H. (1936), 'The role of the lipids in relation to flour quality', *Cereal Chem.*, 13, pp. 318-31.
9. Yenofsky, R.L., Fine, M., and Liu, C. (1988), 'Isolation and characterization of a soybean (Glycine max) lipoxygenase-3-gene', *Mol. Gen. Genet.*, 211, pp. 215-22.

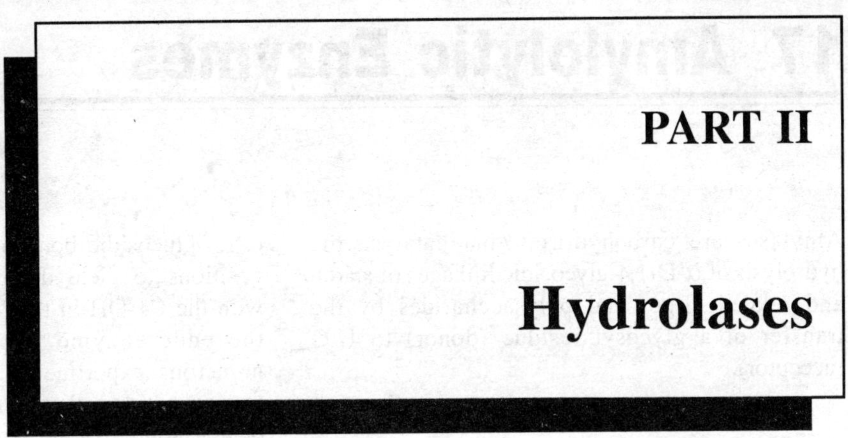

The Hydrolases are a large group of enzymes which have in common the involvement
of water in formation of product.

$$X\text{-}Y + H_2O \longrightarrow HX + YOH$$

Important subgroups include carbohydrases and proteases.

Carbohydrases:

    A.   Amylolytic enzymes
    B.   Pectic enzymes
    C.   Cellulolytic enzymes

Proteases:

    A.   Cysteine proteases
    B.   Aspartic proteases
    C.   Serine proteases
    D.   Metalloproteases

17 Amylolytic Enzymes

Amylases are carbohydrases that catalyse the hydrolysis of α-D-1,4-glycosidic linkages of starch and related oligo- and polysaccharides by the transfer of a glycosyl residue (donor) to H_2O (acceptor).

17.1 α-AMYLASE

α-Amylase (EC. 3.2.1.1, 1,4-α-D-glucan glucano-hydrolase) is an endo-glucosidase that cleaves the α-1,4-glucosidic bond of the substrate at internal positions to yield dextrins and oligosaccharides with the C1-OH in the α-configuration. Although the endo-enzyme implies random cleavage, numerous experiments have suggested that the enzyme action follows a definite pattern depending on the source of the enzyme. An α-1,4 linkage neighbouring an α-1,6 branching point in the substrate is resistant to attack by the enzyme. α-Amylase is found almost universally distributed in plants, animals, bacteria, and fungi.

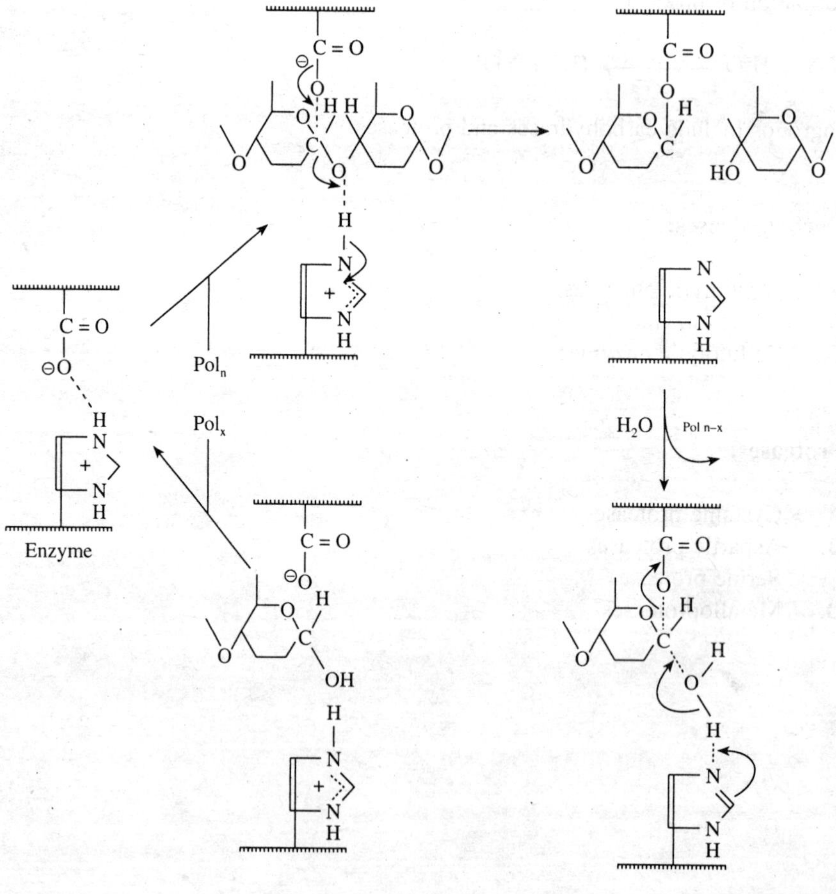

Fig. 17.1 *A proposed mechanism for α-amylase.*

The *Aspergillus oryzae* enzyme consists of 478 amino acid residues folded into two domains. Domain A, which contains the N-terminal 380 amino acid residues, consists of a $(\beta/\alpha)_8$ barrel with eight β-strands alternating with eight helices joined by loops. The loop linking the β strand 3 and helix 3 contains an extended chain of three β strands. Domain B is an 8-stranded antiparallel β sheet linked to domain A via a single polypeptide chain. Four disulphide bonds are located at residues 30-38, 150-164, 240-283, and 439-474. The enzyme is N-glycosylated at Asn 197.

The barley isoenzyme 2 shows the same basic features in the overall structure. Domain A (1-88, 153-350) forms a $(\beta/\alpha)_8$ barrel. The loop protruding between β strand 3 and α helix 3 of domain A constitutes domain B, which assumes an irregular fold stabilized by Ca^{2+} ions. The C-terminal domain C (351-403) forms a 5-stranded antiparallel β sheet.

The active side of the two α-amylases described is located in a cleft at the C-terminal side of the (strands of the $(\beta/\alpha)_8$) barrel of domain A. The two catalytic amino acid residues in the *Aspergillus oryzae* α-amylase are Glu 230 and Asp 297.

17.1.1 Reaction Mechanism

As shown in figure 17.1, the enzyme contains a carboxyl and a nitrogen group (imidazole group) in the active site. The substrate forms an adsorptive complex with the enzyme which positions the susceptible glucosidic bond in juxtaposition with the carboxyl anion and imidazolium group. In the proposed scheme the carboxyl anion serves as an attacking nucleophile on the C (1) position of the substrate and this attack is aided by protonation of the linkage by the general acid (imidazolium ion). A glucosyl-enzyme intermediate involving covalent bonds is formed as a result. In the deglucosylation reaction, the imidazole group (unprotonated) assists as a general base to remove a proton from water leaving the OH^- to attack at the C(1) position of the glucosyl-enzyme. The configuration about C(1) is not changed by this double displacement mechanism so the product of the reaction is of α-configuration.

17.1.2 Optimum pH and Temperature

The pH optimum varies depending on the enzyme source (6.0-7.0 for mammalian, 4.8-5.8 for *Aspergillus oryzae*, 5.86-6.0 for *Bacillus subtilis*, 5.5-7.0 for *Bacillus licheniformis*). Temperature optimum for activity varies 70-72°C for α-amylase produced by *Bacillus subtilis*, and 90°C for the enzyme from *Bacillus licheniformis*. The applications of α-amylase in starch processing are at temperatures ≥ 105°C.

17.2 β-AMYLASE

β-Amylase (EC 3.2.1.2, 1,4-α-D glucan maltohydrolase) occurs commonly in seeds of higher plants and in sweet potatoes. It is an exoenzyme that successively cleaves maltosyl units from the non-reducing end of the polymer substrate to yield maltose with the C1-OH in the β-configuration.

Plant unlike α-amylase, which bypasses the α-1,6 branch point, β-amylase stops the action at such locations. Plant β-amylases often consists of several isoforms. Soybean β-amylase exists predominantly as isoenzymes 2 and 4 with pI 5.25 and 5.5 respectively. The maize enzyme comprises two isoenzymes with pI of 4.25 and 4.40. Barley and wheat consist of two isoenzymes, free and bound forms. The bound form is converted to the free form during germination. β-amylase has also been found in several species of *Bacillus*.

The amino acid sequences are known for β-amylases from soybean, sweet potato, barley and rye. Sweet potato β-amylase is homotetrameric, whereas most plant and bacterial enzymes are monomeric enzymes. The sweet potato enzyme (subunit) and the seed enzymes of soybean, barley, and rye share ~ 50-60 per cent similarities in amino acid sequences. The subunit of sweet potato mature enzyme is composed of 498 amino acids with a calculated MW = 55,880, compared to 495 for soybean, and 535 for barley.

The crystal structure of the soybean β-amylase (isoenzyme 2) is composed of a large domain and a small domain, separated by a long cleft. The large domain consists of a $(\beta/\alpha)_8$ structure with the parallel β strands wound into a barrel, each β strand alternating with an α-helix. The active site is situated in a cleft at the interface between the two domains. The catalytic Glu 186, identified by affinity labeling is located at the base of the active site and is included in the highly conserved regions among both plant and bacterial β-amylases.

17.2.1 Reaction Mechanism

In developing a mechanism for the action of β-amylase one must explain why the enzyme splits off only maltose units and why there is inversion of configuration around C(1) of the hydrolysed bond. It is postulated that the enzyme has at least three specific groups, X, A and B in the active site which are involved in binding and transformation of substrate. The X group recognizes the C(4)—OH group at the non reducing end of the polysaccharide chain. When there is proper interaction between X and OH of C(4) the second glucosidic linkage of the substrate will be in proper juxtaposition with respect to the catalytic groups A and B.

The mechanism for inversion of configuration by β-amylase may be postulated to occur as shown in figure 17.2. The proposed mechanism takes into account the four groups experimentally demonstrated to be in the active site. After formation of the enzyme-substrate complex, a nucleophilic attack of the sulphydryl group C(1) is facilitated by the carboxylate group acting as a general base while the imidazolium group acts as a general acid to donate a hydrogen to the glycosidic oxygen. This leads to formation of maltosyl-enzyme intermediate. In hydrolysis of the intermediate, the carboxylate group acts as a general base to facilitate a backside attack of water at C(1) to release β-maltose and regenerate the enzyme. Group X serves to position the substrate properly at the active site.

17.2.2 Optimum pH

The pH optimum for activity ranges from 5.0 for wheat, malt and sweet potato to 6.0 for soybean and pea.

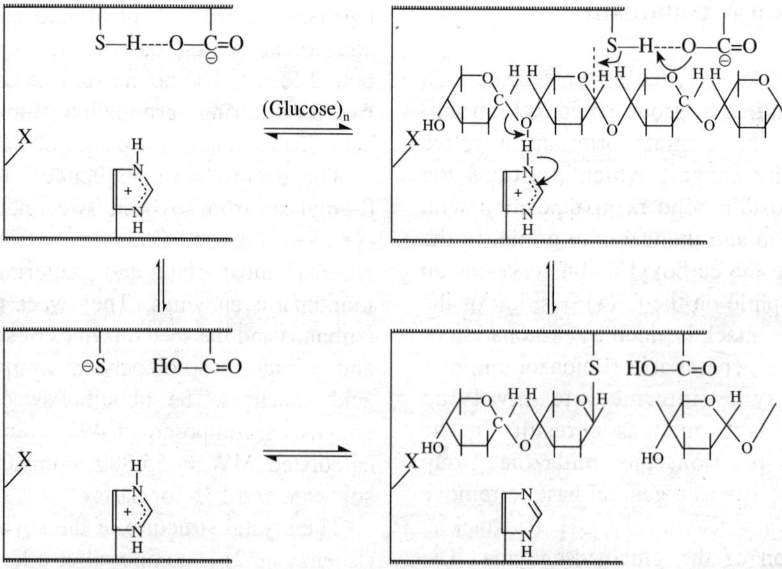

Fig. 17.2 *A proposed mechanism of β-amylase-catalysed hydrolysis of amylose.*

17.3 GLUCOAMYLASE

Glucoamylase, also occasionally called γ-amylase or amyloglucosidase in some literature, is an exoglucosidase that catalyses the hydrolysis of both α-D-1,4- and α-D-1,6-glucosidic linkages at the branch point, although hydrolysis of the latter occurs at a slower rate. The enzyme removes glucose units successively from the surrounding end of the substrate. The rate of hydrolysis increases with the chain length of the substrate. The end product is exclusively glucose in the β-configuration. Glucoamylase is a microbial enzyme found only in molds and yeasts. The enzyme is commercially produced from *Aspergillus niger*.

Glucoamylase is a glycoprotein containing 5-20 per cent carbohydrate mainly, mannose, glucose, galactose and glucosamine. The molecular weight varies depending on the source of the enzyme, and may exist in multiple forms. Fungal glucoamylase generally exists in multiple forms. For example, *Aspergillus niger* glucoamylase exists in two forms, GAI and GAII. The GAI form (616 amino acids, MW 71KD) consists of a catalytic site region (1-440), a Thr- and Ser-rich linker region (441-512) heavily glycosylated, and a C-terminal region (513-616). The latter is missing in the GAII form (MW 61 KD), and is referred to as the granular starch binding domain.

17.3.1 Optimum pH and Temperature

The enzyme has a pH optimum of 4.0-4.4, with stability over a pH range of 3.5-5.5. The enzyme is stable at a temperature range of 40-65°C with optimum range 58-65°C in the hydrolysis of starch at pH 4.2.

17.4 ACTIVITY ASSAY

Four general assay procedures have been used for studying the action of amylases on starch. During hydrolysis there is (a) decrease in viscosity, (b) loss in ability to give a blue colour with iodine, (c) an appearance of reducing groups, and (d) an increase in maltose, glucose, or dextrins. The three types of amylases can be distinguished from each other on the basis of two or more of these criteria (Table 17.1).

α-amlyase, because it hydrolyses the α-1,4 glucosidic linkages of the polysaccharide at random, will cause a very rapid loss in viscosity and iodine colour-forming ability because both of these parameters are dependent on the integrity of the large polymer. Hydrolysis of one or a very few glucosidic bonds near the centre of the substrate will cause a marked change in these parameters while hydrolysis of one or a very few bonds near the end of the polysaccharide chains by the exo-splitting enzymes, β-amylase and glucoamylase, will have little effect.

TABLE 17.1

Criterion	Relative rates		
	α-Amlyase	β-Amylase	Glucoamylase
Reducing group formation	Fixed as equal	Fixed as equal	Fixed as equal
Loss in viscosity	Fast	Slow	Slow
Loss in iodine colour	Fast	Slow	Slow
Maltose production	Slow	Fast	None
Glucose production	None	None	Fast

The rate of decrease in viscosity or iodine colour cannot be used alone to differentiate among the three enzymes because the rate of change in these parameters is dependent not only upon the type of enzyme but also on the concentration of that enzyme. A very high concentration of the exo-splitting enzymes will cause a rapid loss in viscosity just because there are so many bonds hydrolysed rapidly. Conversely, a low concentration of α-amlyase will cause a slow change in these parameters. Therefore, the rate of change in the number of reducing groups formed must be determined as a basis of comparison. Any of the methods, including those which use 3,5-dinitrosalicylic acid, potassium ferricyanide and alkaline cupric solutions may be used to determine the number of reducing groups formed.

17.5 APPLICATIONS OF AMYLASES IN FOOD PROCESSING

The classical uses of amylases and also some emerging applications that may be of importance in the future include:

17.5.1 Starch Processing

The food industry uses enzymes as processing aids to convert starch bearing materials to starches, starch derivatives and starch saccharification products.

Industrial processes for starch hydrolysis to glucose rely on inorganic acid or enzyme catalysis. The use of enzymes is preferred currently and offers a number of advantages associated with improved yields and favourable economics. Enzymatic hydrolysis allows greater control over amylolysis, the specificity of the reaction and the stability of the generated products. The milder reaction conditions involve lower temperature and near neutral pH, thus reducing unwanted side reactions. Fewer off-flavour and off-colour compounds are produced, especially 5-hydroxy-2-methylfurfuraldehyde, anhydroglucose compound, and undesirable salts. Enzymatic methods are favoured because they also lower energy requirements and eliminate neutralization steps.

Starch derivatives obtained via the liquefaction process have commercial value as food ingredients. The products have a DE value of about 10 and are sold as maltodextrins, which are carbohydrate preparations composed of 3 per cent maltose, 4 per cent maltriose and 93 per cent tetroses or larger polysaccharides. Maltodextrins are valuable to the food industry since they serve as thickening agents and additives able to promote drying of hygroscopic food components.

17.5.1.1 Gelatinization and liquefaction of starch

Current starch processing methods require the gelatinization of corn starch after wet milling. This pretreatment is essential before an extended period of enzymatic hydrolysis. The starch which is present as $5\text{-}25\mu$ m particles, is exposed to temperatures above 60°C that swell and disrupt the granules. The addition of heat-stable α-amylase from strains of *Bacillus subtilis* at this stage causes random hydrolysis of glycosidic bonds and significantly thins the starch slurry. The starch slurry so produced is of manageable viscosity.

After gelation of the starch, process of liquefaction begins. The thinned starch slurry composed of 30-40 per cent solids is treated with a thermostable α-amylase able to function at pH 6.5. The starch is contained in a steam jet cooker operated in a continuous mode for 5-10 min. at 103-107°C. Afterward, a 1-2 hr treatment at lower temperature of about 95°C permits hydrolysis of the thinned starch to a syrup of 0.5-1.5 DE (dextrose equivalent). Digestion is allowed to continue until a DE of 10-15 is attained. The net yield of dextrose after complete enzymatic digestion of the corn starch with bacterial amylase is 95-97 per cent if hydrolysis is coupled with a fungal amylase treatment.

17.5.1.2 Saccharification of starch

The saccharification of starch, that is, the hydrolysis of starch to glucose, is accomplished with fungal glucoamylse. The enzyme is effective since it has both exoamylase and debranching activity and is important for the production of a corn syrup of high DE or of crystalline dextrose. A debranching enzyme such as pullulanase may also be used. Fungal amylase can hydrolyse starch to glucose almost completely after a treatment time of 3-4 days at pH 4.0 and 60°C. The low pH optimum of fungal amylolytic enzymes permits the convenient use of acid conditions for the saccharification. Such conditions reduce unwanted isomerization reactions to fructose and other sugars that may reduce the glucose yield. Moreover, acid conditions restrict the growth of contaminating micro-organisms in the saccharification reactors.

17.5.2 Baking

The main reasons for the supplementation of flour with amylases include:

1. Amylases increase the level of fermentable sugars present in doughs. Poor loaf volume would otherwise result.
2. Amylases improve crust colour. The reducing sugars produced by amylases react with other components in bread to give Maillard reaction products. These are responsible for the typical golden colour of the crust.
3. The flavour of bread is improved both by the simple sugars produced by amylases and by these Maillard reaction products.
4. Gas retention properties of the dough are improved by starch modification resulting from amylases.
5. The crumb has improved moisture retention properties also as a result of starch modification by amylases.
6. Heat-stable amylases retard the staling of bread.

Little is known about the mechanisms by which amylases improve the gas retention properties of the dough and the moisture retention properties of the crumb. The mechanism by which heat-stable amylases exert an anti-staling effect is not well understood, but several products are in the market for this purpose. The other three effects attributed to amylase addition are all related to their ability

to produce simple sugars in bread which are both fermentable and reducing.

The level of fermentable sugars present in flour is quite low, less than 0.5 per cent. The level is too low to support yeast metabolism to the extent to obtain optimum gassing power (CO_2 production) from the yeast. Therefore poor loaf volume results. Both β-amylase and glucoamylase produce sugars that yeast can ferment. Glucoamylase can be added for this purpose, but its addition to bread usually accomplishes other ends as well. β-amylase is inherent in normal flour at levels high enough that further supplementation offers no advantage. However, without supplementation of the dough with other enzymes, β-amylase by itself will not increase fermentable sugars to the levels desired, i.e. loaf volume is not optimum. The reason is that during the fermentation and proofing stages of dough handling, only small percentage of damaged starch present in the flour is accessible to enzyme attack; substrate availability is the rate-limiting factor. α-amylase will produce many small dextrins on which the β-amylase can more readily act. Thus, although the dextrins produced by α-amylase are not themselves fermentable by yeast, it is α-amylase that limits the production of fermentable sugars. α-amylase is not present in sufficient quantities in normal flour to adequately fill this need. While β-amylase is the enzyme specifically responsible for increasing the level of fermentable sugars, it is not the limiting factor in the rate of fermentable sugar formation.

Thus the major direct effect of α-amylase supplementation in doughs, along with the inherent β-amylase, is an increase in the fermentable sugar level. This results in greater gas production by the yeast and translates to increased loaf volume.

The sugars produced by these enzymes are not only fermentable, but reducing as well. Higher concentrations of reducing sugars will result in increased levels of Maillard reaction products, which are responsible for the darkening of the crust. Along with the general maturing of the dough associated with better yeast growth, Maillard reaction products also give bread much of its typical flavour.

Glucoamylase is not widely used in baking. It does increase the fermentable sugar level in the dough but, as discussed above, β-amylase is sufficient for that and does not need to be added. Glucoamylase is used in bread because it produces glucose, which has a greater perceived sweetness than does maltose, the product of β-amylase action. Glucoamylase also exhibits a relatively high thermostability. It will remain active after the yeast have been killed and are unable to consume the produced glucose. As a result, it is possible to use glucose in a baked product, reduce the amount of added sugar, and still have the desired colour and sweetness.

17.5.2.1 Amylases as anti-staling agent

During the aging of baked products many physical and chemical changes can occur to the flour components. However, it is the change in physical form of the starch fraction that is considered to be mainly responsible for the staling process.

As applied to baked products such as bread, staling is thought to be due to the slow passage of water from the starch to other components of the bread.

During the dough stages of baking, most of the starch in the flour is in semicrystalline granules. As higher temperatures are reached in the oven, the granular starch begins to gelatinize to absorb water, swell and lose crystallinity. As the granules begin to rupture, much of the highly soluble amylose is leached out of the granule into the open matrix of the bread. Within the remnants of the starch granule is a mass of partially swollen amylopectin. This amylopectin core exists because not enough water and energy are available for complete solvation. After baking, as the bread cools, the dissolved amylose retrogrades or recrystallize within a few hours. This is an intermolecular association in which the long, linear amylose chains form an ordered and very stable array. After this initial rapid retrogradation of the amylose, a much slower rate of retrogradation of the amylopectin occurs which centres around the intramolecular ordering of the outer amylopectin branches. At the same time, cross-links develop between the retrograded amylose in the matrix of the bread and the remaining granules. This results in an overall rigidity that is detected as crumb firmness.

The amylases have a significant anti-staling effect on the structure of starch in the final product. Several models exist for explaining the mechanism for the anti-staling effect of amylases in bread. The simplest theory is that the soluble saccharides produced by the amylases themselves exert a staling effect since rate of retrogradation is dependent on the size of the starch or starch fragments and optimum chain length for retrogradation of amylose is of the order of 500-1,000 glucose units. Low molecular weight saccharides (> 10 glucose units) favour interaction with water rather than reassociation with each other or larger starch chains.

SUMMARY

Amylases are carbohydrases that catalyse the hydrolysis of α-D-1,4 glycosidic linkages of starch and related oligo- and polysaccharides by the transfer of a glycosyl residue (donor) to H_2O (acceptor).

α-Amylase (EC. 3.2.1.1, 1,4-α-D-glucan glucanohydrolase) is an endoglucosidase that cleaves the α-1,4-glucosidic bond of the substrate at internal positions to yield dextrins and oligosaccharides with the C1-OH in the α-configuration. α-Amylase is found almost universally distributed in plants, animals, bacteria and fungi. The optimum pH and temperature varies depending on the enzyme source.

β-Amylase (EC. 3.2.1.2, 1,4-α-D glucan maltohydrolase) occurs commonly in seeds of higher plants and in sweet potatoes. It is an exoenzyme that successively cleaves maltosyl units from the non-reducing end of the polymer substrate to yield maltose with the C1-OH in the β-configuration. Unlike α-amylase, which bypasses the α-1,6 branch point, β-amylase stops the action at such locations. Plant β-amylases often consist of several isoforms. The pH optimum for activity ranges from 5.0 for wheat, malt and sweet potato to 6.0 for soybean and pea.

Glucoamylase, also occasionally called γ-amylase or amyloglucosidase in some literature, is an exoglucosidase that catalyses the hydrolysis of both α-D-1,4- and α-D-1,6 glucosidic linkages at the branch point, although hydrolysis of the latter occurs at a slower rate. The enzyme removes glucose units successively from the surrounding end of the substrate. The end product is exclusively glucose in the β-configuration. Glucoamylase is a microbial enzyme found only in moulds and yeasts. The enzyme is commercially produced from *Apergillus niger*. The enzyme has a pH optimum of 4.0-4.4, with stability over a pH range of 3.5-5.5. The enzyme is stable at a temperature range of 40-65°C with optimum range 58-65°C in the hydrolysis of starch at pH 4.2.

Four general assay procedures have been used for studying the action of amylases on starch. During hydrolysis there is (a) a decrease in viscosity, (b) a loss in ability to give a blue colour with iodine, (c) an appearance of reducing groups, and (d) an increase in maltose, glucose or dextrins.

The classical uses of amylases and also some emerging applications that may be of importance in the future include gelatinization and liquefaction of starch, saccharification of starch and as anti-staling agent in baking.

RECOMMENDATIONS

1. Amos, A.J. (1955), 'The use of enzymes in baking industry', *J. Sci. Food Agric.*, 6, pp. 489-95.
2. Anon. (1984), A. DIAZYME L200, fungal glucoamylase for starch hydrolysis, Miles Laboratories, Inc., *Enzyme Products*, Elkhart, IN.
3. Anon (1984), B. KINASE-HT, thermal stable bacterial alpha-amylase for the brewing industry, Miles Laboratories, Inc., Biotech Division, Elkhart, IN.
4. Barrett, F.F. (1975), 'Enzyme uses in the milling and baking industries', *Enzymes in Food Processing* (G. Reed, edn.), 2nd edn., pp. 301-30, Academic Press, New York.
5. Ji, E. S., Mikami, B., Kim, J.I., and Morita, Y. (1990), 'Positions of substituted amino acids in soybean α-amylase isoenzymes', *Agric. Biol. Chem.*, 54, pp. 3065-7.
6. Kadziola, A., Abe, J.I., Svensson, B., and Haser, R. (1994), 'Crystal and molecular structure of barley α-amylase', *J. Mol. Biol.*, 239, pp. 104-21.
7. Kreis, M., Williamsson, M., Buxton, B., Pywell, J., Heijgaard, J., and Svendsen, I. (1987), 'Primary structure and differential expression of α-amylase in normal and mutant barleys', *Eur. J. Biochem.*, 169, pp. 517-25.
8. Mac Allister, R.V., Wardrip, E.K., and Schnyder, B.J. (1975), 'Modified starches, corn syrups containing glucose and maltose, corn syrups containing glucose and fructose, and crystalline dextrose', *Enzymes in Food Processing* (G. Reed, edn.), 2nd edn., pp. 331-59, Academic Press, New York.
9. Mikami, B., Nomura, K., Majima, K., and Morita, Y. (1989), 'Structure of soybean α-amylase and the reactivity of its sulphhydryl groups', *Denpun Kagaku* (Japan), 36, pp. 67-72.
10. Readley, J.A. (1976), 'Starch Production Technology', *Applied Science*, London.
11. Takeda, Y., and Hizukuri, S. (1981), 'Re-examination of the action of sweet-potato beta-amylase on phosphorylated α-1-4 α-D-glucan', *Carbohydr. Res.*, 89, pp. 174-8.
12. Thoma, J.A, Spradlin, J.E., and Dygert, S. (1971), 'Plant and animal amylases', *The Enzymes*, 5, pp. 115-89.
13. Toda, H., Nitta, Y., Asanami, S., Kim, J.P., and Sakiyama, F. (1993), 'Sweet potato β-amylase, Primary structure and identification of the active-site glutamyl residue', *Eur. J. Biochem.*, 216, pp. 25-38

18 Pectic Enzymes

Pectic enzymes constitute a unique group of enzymes that catalyse the degradation of pectic polymers in plant cell walls. Pectic substances are heterogeneous in chemical structure and molecular size. The basic chemical structure of pectin is α-D-galacturonan or α-D-galacturonoglycan, a linear chain of α-D-galactopyranosyluronic acid units. The polymer contains varying degrees of esterification, with certain carboxyl groups esterified with methoxyl groups. In some cases, such as beet pectin, esterification with acetyl groups occurs. The backbone chain is usually interrupted by L-rhamnose linked α-1,2 and β-1,4 to the preceding and succeeding galacturonic acid respectively. Branching also occurs with neutral sugars, forming side chains of α-1,5-arabinan and β-1,4-galactan.

18.1 CLASSIFICATION OF PECTIC SUBSTANCES

Pectic substances can be classified according to the modification of the backbone chain.

1. Pectic acids: Galacturonans that contain negligible amounts of methoxyl groups. Pectates are salts of pectic acids.
2. Pectinic acids: Galacturonans with various amounts of methoxyl groups. Pectinates are salts of pectinic acids.
3. Pectins: A generic name for mixtures of widely differing composition containing pectinic acid as the major component. Pectins in native form as located in the cell wall may be interlinked with other structural polysaccharides and proteins to form insoluble propectin. Solubility of pectins can be achieved by degradation such as heating in acid medium. The soluble pectin thus obtained is partially degraded and heterogeneous.

18.2 CLASSIFICATION OF PECTIC ENZYMES

Pectic enzymes can be classified according to their mode of action.

1. Polygalacturonase (PG): Catalyses the hydrolytic cleavage of α-1,4 glycosidic bonds. The exo-PG (exo-poly 1,4-α-D-galacturonide) galacturonohydrolase, (EC. 3.2.1.67) cleaves from the non-reducing end, and the endo-PG (endo-poly 1,4-α-D-galacturonide) glycanohydrolase, (EC. 3.2.1.15) attacks the substrate randomly.
2. Pectinesterase (PE) (Pectin pectylhydrolase, EC. 3.1.1.11): Catalyses the hydrolysis of the methyl ester groups, resulting in the deesterification of pectin. The enzyme attacks preferentially on a methyl ester group of a galacturonate unit next to a non-esterified galacturonate unit.
3. Pectate lyase (PEL): Catalyses the cleavage of non-esterified galacturonate units via α-elimination. Both exo-PEL (exo-poly 1,4-α-D-galacturonide lyase, EC. 4.2.2.9) and endo-PEL (endo-poly 1,4-α-D-galacturonide lyase, EC. 4.2.2.2) enzymes exist. Pectate and low-methoxyl pectin are the preferred substrates for the enzymes.
4. Pectin lyase (PNL): Catalyses cleavage of esterified galacturonate units by β-elimination. All PNLs studied are endo-enzymes.

18.3 POLYGALACTURONASE (PG)

The PG hydrolyses glycosidic linkages in pectic substances by the use of water.

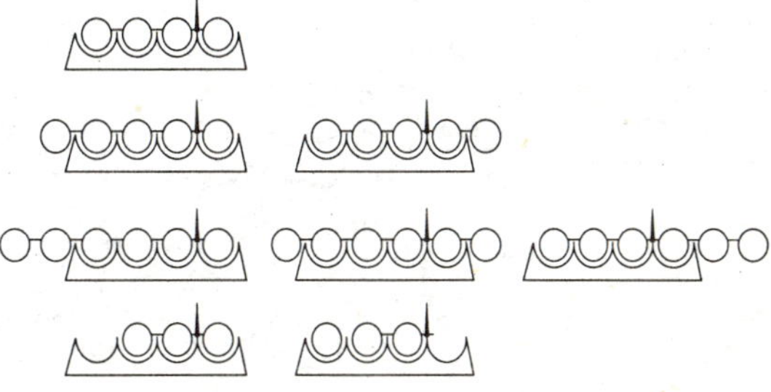

Several methods are available to distinguish between the exo-and endo-splitting enzymes. One of the most useful methods is to compare the rate of decrease in viscosity with rate of hydrolysis as measured by increase in reducing groups. An endo-splitting enzyme causes ~ 50 per cent reduction in viscosity when only 3 to 5 per cent of the glycosidic bonds have been hydrolysed, while with an exo-splitting enzyme ~ 10 to 15 per cent hydrolysis of glycosidic bonds are needed to produce a 50 per cent reduction in viscosity.

A second method which will distinguish between an endo- and an exo- splitting enzyme is the nature of the products formed near the beginning of the reaction. No low molecular weight products will be found near the beginning of reaction with an endo-splitting enzyme while with an exo-splitting enzyme low molecular weight products will be present near the very beginning.

PG largely comes from microbial sources. PG is commonly found in the extracellular secretions of pathogenic species of fungi and bacteria. PG in higher plants has been extensively studied in ripe tomato fruit. Tomato PG exists in two forms; both are endo-enzymes. PG1 has a MW of 80 KD and is 50 per cent inactivated at 78°C. PG2 has a MW of 44 KD and is 50 per cent inactivated at 57°C. PG1 has an optimum stability at pH 4.3, whereas PG2 is most stable at pH 5.6. A single PG gene codes for each of the recognized isoenzymes, consistent with the conclusion that PG1 and PG2 consist of the same catalytic polypeptide, and that PG1 is formed by the association of PG2 with a β subunit. β subunit (heat stable non-dialyzable factor) is a glycoprotein immunologically distinct from the PG polypeptide. PG1 is made up of one molecule of β subunit and one molecule of PG2. The β subunit is non-catalytic; only the PG polypeptide has enzyme activity.

The enzyme activity in the early stages of ripening in mainly due to PG1. PG1 continues to increase during ripening, and is detected at all stages of ripening. The lower-MW and less heat-stable PG2 appears late in the process, and becomes the major enzyme at the ripening stage. This apparent sequential appearance of the two enzymes is the result of the regulatory action of the β subunit, which is found at low levels in green tomatoes, and in increasing amounts with ripening. It is now generally agreed that PG2 is the only endogenous PG in tomatoes, and that PG1 is formed during ripening as β subunit is produced to react with PG2.

18.3.1 Mechanisms and Action Pattern

Exo-PG hydrolyses the non-reducing end unit of a polygalacturonic chain, giving galacturonic acid as a major reaction product. Depolymerization is interrupted by branching that occurs in the substrate. The rate of hydrolysis increases with substrate size, and reaches a maximum with a degree of polymerization of 20 for carrot and peach exo-PGs. The specificity and action pattern of endo-PG are determined by the nature of the active site. The rate of hydrolysis by yeast endo-PG decreases with a shortening of the substrate. The binding site of *Aspergillus niger* endo-PG is comprised of four subsites and cleavage occurs between subsites 1 and 2 (Figure 18.1). A tetramer substrate undergoes (3+1) cleavage into trigalacturonic and galacturonic acid. A pentamer gives two productive complexes, both satisfying the complete occupancy of all four sites, yielding a (4+1) and (3+2) cleavage. Likewise, for a hexagalacturonic acid substrate, three productive complexes can be envisioned: (5+1), (4+2), and (3+3).

Fig. 18.1 *Action pattern of endo-polygalacturonases from Aspergillus niger.*

The role of NaCl, needed for maximum activity of the PG is not clear but it appears to function in part to prevent inhibition of the enzyme by the products. Some of the PGs require Ca^{2+}. Whether Ca^{2+} performs a role in binding and / or catalysis, in maintaining the conformation of the enzyme or whether it masks the carboxylate groups of the substrate is not clear.

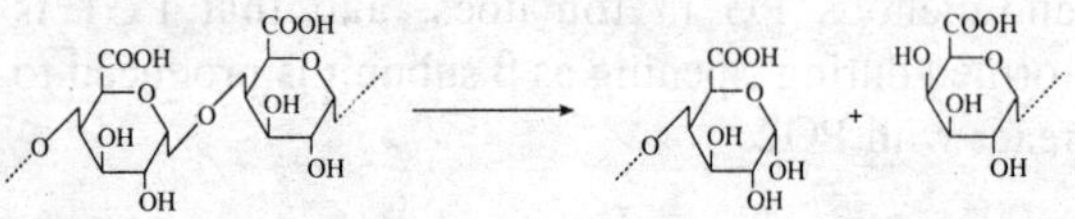

Fig. 18.2 *Depolymerization action of polygalacturonase.*

The pH optima for most of the PG enzymes are in the range of 4.5 to 6.0.

18.3.2 Activity Assay

Activity of the PGs may be followed by measuring
 (a) the rate of decrease in viscosity of the reaction mixture,
 (b) the rate of formation of reducing groups,
 (c) the decrease in optical rotation, or
 (d) the decrease in precipitability by calcium ions or non-polar solvents.

18.4 PECTINESTERASE (PE)

PE removes methoxyl groups from methylated pectic substances and therefore belongs to the subdivision of enzymes which hydrolyse carboxylic acid esters.

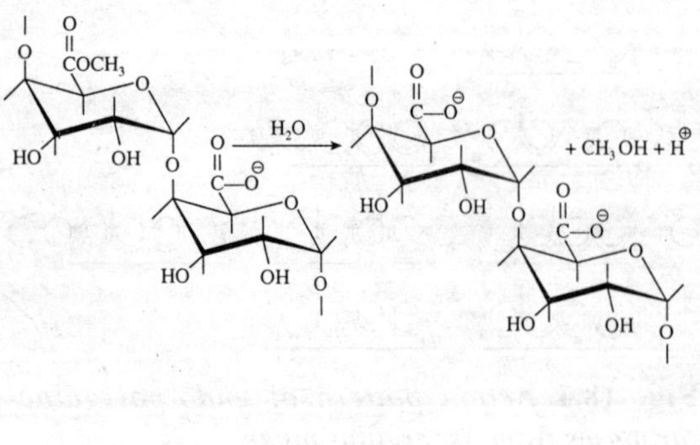

PE is widely present in various fruit plants. The enzyme usually exists in multiple forms and is found in cell wall fractions. Fungal PEs have also been isolated as extracellular enzymes.

Tomato fruit contains at least two PE isoenzymes. Both isoenzymes PE1 and PE2 increase during the initial phase of ripening. After the breaker stage, there is a decrease in PE1, but an increasing accumulation of PE2 until the full colour stage of development. PE2 has a MW of 23 KD and a pH optimum of 7.6. Soybean PE is a 33 KD protein with maximum activity at pH close to 8.0. Two PE isoenzymes are found in banana pulp; both having the same MW of 30 KD, but different PIs of 8.8 and 9.3. The PEs show optimum activity at pH 7.5. Orange PE is another well-studied enzyme. Two isoenzymes, PE1 and PE2, have been purified; both are 36.2 KD proteins, but with different PIs of 10.05 and 11.0, respectively. Grapefruit pulp contains two isoenzymes with MW of 51KD and 36 KD. Apple and kiwi both contain two PE isoenzymes.

The primary structure of tomato PE is known. Protein sequencing identifies 305 amino acid residues with a MW of 33,239. The enzyme contains two disulphide bridges (Cys 98-Cys 125, and Cys 166- Cys 200).

18.4.1 Mechanism of Demethylation

PE removes methoxyl groups in pectin by a nucleophilic attack of the enzyme on the ester, resulting in the formation of an acyl-enzyme intermediate with the release of a methanol. This is then followed by deacylation, which is the hydrolysis of the intermediate to regenerate the enzyme and a carboxylic acid (Figure 18.3).

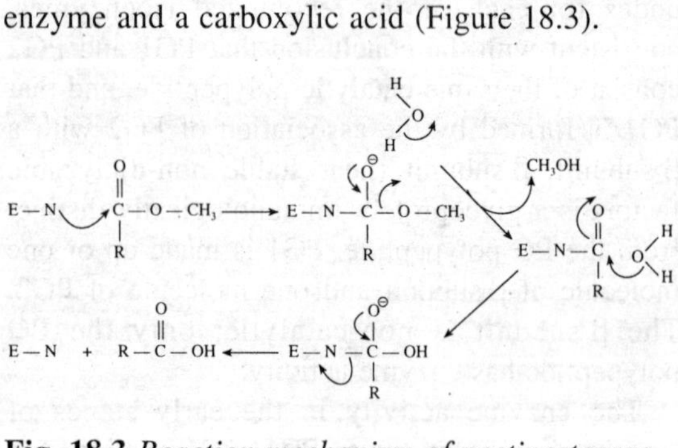

Fig. 18.3 *Reaction mechanism of pectinesterase.*

18.4.2 Activity Assay

Activity of the enzyme on pectin may be followed most conveniently and continuously in a pH meter at pH 7.5 since a proton is released when the ester bond is hydrolysed. The rate of methanol production and the Ca^{2+} precipitability of the pectic acid formed have also been used to determine activity of the enzyme.

18.5 PECTATE LYASE (PEL)

PEL split the glycosidic bond by trans elimination of hydrogen from the C 4 and C 5 positions of the aglycone portion of the substrate. Therefore, these enzymes belong to the lyase group of enzymes.

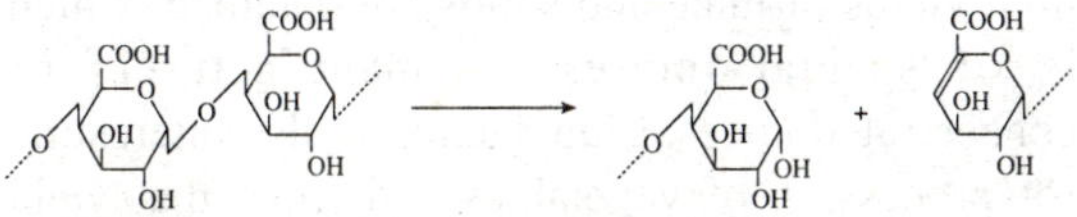

Pectate lyases (PELs) are microbial extracellular enzymes. Enzymes of the genera *Erwina* and *Bacillus* are among the best known to produce soft-rot symptoms in plants, although the enzyme is also found in other micro-organisms.

Erwina chrysanthemi produces pectate lyases in isoforms that can be grouped according to their pIs: acidic (pH 4-5), neutral (pH 7-8.5), and alkaline (pH 9-10). The number of isoenzymes in each group may vary depending on the strains. Most strains studied produce five isoenzymes— one acidic (PELA), two neutral (PELB and C), and two alkaline (PELD and E). PELA (pI 4.2) contains 361 amino acids with a molecular weight of 38,756. PELB (pI 8.8) consists of 392 amino acids with a molecular weight of 37,922. The amino acid sequence predicted for PELC contains 375 amino acid residues. The MW is estimated to be 37,676. The PELE protein has 355 amino acids and a calculated MW of 38,078. All isoenzymes are activated by Ca^{++}. All the *Bacillus* PELs exhibit endo-action and require Ca^{++} for activation. A purified PEL excreted from *Bacillus subtilis* strain S0113 has a pI of 9.6 and a MW of 42 KD. The enzyme has optimum activity at pH 8.4 and 42°C. As in all other PELs, the enzyme is activated by Ca^{++}.

18.5.1 Reaction Mechanism

It is known that in alkaline medium, pectin undergoes deesterification, accompanied by degradation via β-elimination. Similar splitting of glycosidic bonds also occurs in neutral pH solution at elevated temperature. Pectate, which consists mostly of carboxyl groups, is relatively resistant. These observations are likely to be related to the electron-withdrawing methyl ester that renders the C5-H increasingly acidic. Abstraction of the proton at C5 by hydroxyl ions results in the formation of a carbanion that is stabilized by the C6 carboxyl group. This step is then followed by elimination at C4 with the formation of a double bond between C4 and C5 (Figure 18.4). The rate of elimination of pectin is enhanced by the addition of Ca^{++} and K^{+}. The effect of cation may suggest neutralization of the anionic groups in the pectin chain, thereby increasing the accessibility of hydroxyl ions.

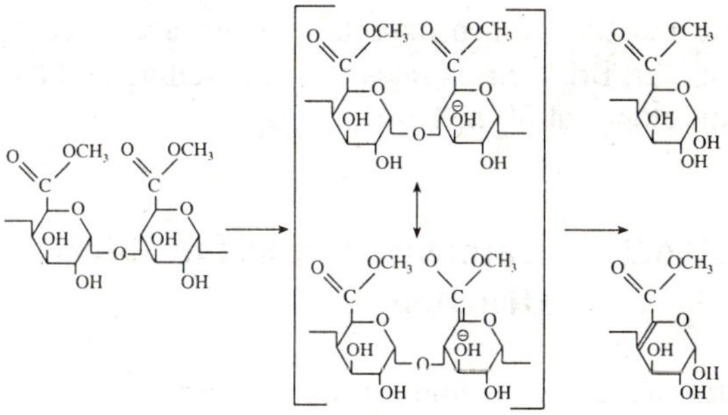

Fig. 18.4 *The proposed reaction mechanism of Pectate lyase.*

18.5.2 Activity Assay

Activity of the trans-eliminases can be measured by the same methods as described for the polygalacturonases. However, it is more specific and convenient to follow the rate of splitting of the glycosidic bond by observing the increase at 235 nm due to formation of the double bond.

18.6 APPLICATIONS IN FOOD PROCESSING

18.6.1 Fruit Juice Clarification

Fruit juice clarification is the oldest and still the largest use of pectinases, applied mainly to deciduous juices and grape juice. The traditional way to prepare such juices is by crushing and pressing the pulp. The raw press juice is a viscous liquid with a persistent cloud of cell wall fragments and complexes of such fragments with cytoplasmic protein. Addition of pectinase lowers the viscosity and causes cloud particles to aggregate to larger units, which sediment and are removed easily by centrifugation or ultrafiltration. Pectinase with a positive surface charge is coated by negatively charged pectin molecules. Partial degradation of this pectin by pectinase results in aggregation of oppositely charged particles. Also, the reduction of the viscosity of the raw juice is due to pectinase action. The clarification of juices by pectin degradation is also important in the manufacture of high Brix concentrates to avoid gelling and the development of the haze.

18.6.2 Treatment of Pulp for Juice Extraction

In the early period of use of pectinase for clarification, it was found, first for blackcurrants, that enzyme treatment of the pulp before pressing improved juice and colour yield. Pulping of such soft fruits, with much more soluble pectin, results in a semigelled mass that is very difficult to press. Enzymatic pectin degradation yields thin free-run juice and a pulp with good pressing characteristics. The better release of anthocyanins of red fruits into the juice, achieved by cell wall destruction, is another advantage of the pulp enzyme process, which has received attention from red wine manufacturers also. The process permits good yields of red grape juice, especially in combination with a mild heat treatment, which then can be fermented directly. This technique is simpler than the classical fermentation on the skins as required for red wine to obtain the desired colour. Enzyme treatment of pulp of olives, palm fruit and coconut to increase oil yield has also been in practice.

18.6.3 Liquefaction

Liquefaction is a process in which pulp is liquefied enzymatically so pressing is not necessary. The enzyme needed is pectinase. The low viscosity values correspond to complete liquefaction, at that stage microscopic examination shows that cell walls disappear. The liquefied juices are almost clear (papaya, cucumber), cloudy (apples, peaches), or pulpy (carrots), depending on the accessibility of the cell wall compounds to the enzymes. The breakdown products increase the soluble solid contents of the juices. In this way, high yields of juice and solids are obtained, which is of special interest to manufacturers of concentrated juices. Liquefaction is also interesting for processing of vegetables and fruits that yield no juice on pressing or for which no presses have been developed (mango, guava, bananas).

18.6.4 Maceration

Maceration is the process by which organized tissue is transformed into a suspension of intact cells, resulting in pulpy products used as base material for pulpy juices and nectars, and as ingredients for dairy products such as puddings and yogurts. Enzymatic degradation of pectin after a mild mechanical treatment often improves product properties if the process is carried out to leave as many cells as possible intact. Since the aim of the enzyme treatment is the transformation of tissue into a suspension of intact cells, pectin degradation should affect only the middle lamella pectin. This process is called enzymatic maceration.

The restricted pectin degradation dissolves, rather than degrades the middle lamella pectin, which may improve the creamy mouthfeel. The process is often used for carrots. A very interesting use of enzymatic maceration is pulp for the production of dried instant potato mash. Enzymatic maceration prevents starch leaking out of the cells, avoiding the quality defect of gluiness of the reconstituted product. Potatoes are blanched to gelatinize starch and to inactivate PE, which in conjunction with the added PG would extend the possibilities for pectin degradation and turn the maceration process into pulp enzyming technology.

SUMMARY

Pectic enzymes constitute a unique group of enzymes that catalyse the degradation of pectic polymers in plant cell walls. Pectic substances are heterogeneous in chemical structure and molecular size and can be classified into:

1. Pectic acids: Galacturonans that contain negligible amounts of methoxyl groups.
2. Pectinic acids: Galacturonans with various amounts of methoxyl groups.
3. Pectins: A generic name for mixtures of widely differing composition containing pectinic acid as the major component.

Pectic enzymes can be classified according to their mode of action.

1. Polygalacturonase (PG) catalyses the hydrolytic cleavage of α-1,4 glycosidic bonds. PG is commonly found in the extracellular secretions of pathogenic species of fungi and bacteria. PG in higher plants has been extensively studied in ripe tomato fruit. The pH optima for most of the PG enzymes are in the range of 4.5 to 6.0. Activity of the PGs may be followed by measuring (a) the rate of decrease in viscosity of the reaction mixture, (b) the rate of formation of reducing groups, (c) the decrease in optical rotation, or (d) the decrease in precipitability by calcium ions or non-polar solvents.
2. Pectinesterase (PE) catalyses the hydrolysis of the methyl ester groups, resulting in the deesterification of pectin. PE is widely present in various fruit plants. Activity of the enzyme on pectin may be followed most conveniently and continuously in a pH meter at pH 7.5 since a proton is released when the ester bond is hydrolysed. The rate of methanol production and the Ca^{2+} percepitability of the

pectic acid formed have also been used to determine activity of the enzyme.

3. Pectate lyase (PEL) catalyses the cleavage of non-esterified galacturonate units via β-elimination. PELs are microbial extracellular enzymes. Activity can be measured by the same methods as described for the polygalacturonases. However, it is more specific and convenient to follow the rate of splitting of the glycosidic bond by observing the increase at 235 nm due to formation of the double bond.
4. Pectin lyase (PNL) catalyses cleavage of esterified galacturonate units by β-elimination.

Fruit juice clarification is the oldest and still the largest use of pectinases.

RECOMMENDATIONS

1. Bauman, J.W. (1981), 'Application of enzymes in fruit juice technology', *Enzymes and Food Processing* (G.G. Birch, N. Blakebrough, and K.J. Parker, eds.), pp. 129-47, Applied Science, London.
2. BeMiller, J.N. (1986), 'An introduction to pectins: Structure and properties, Chemistry and functions of pectins', M.L Fishman and J.J. Jen, eds., *ACS Sym. Ser.*, 310, American Chemical Society, Washington, D.C.
3. Fogarty, W.M., and Ward, O.P. (1974), 'Pectinases and pectic polysaccharides', *Prog. Ind. Microbiol.*, 13, pp. 59-113.
4. Hasegawa, S., and Nagel, C.W. (1962), 'The characterization of an α, β-unsaturated digalacturonic acid', *J. Biol. Chem.*, 237, pp. 619-21.
5. Lee, M., and Macmillan, J.D. (1968), 'Mode of action of pectic enzymes. I. purification and certain properties of tomato pectinesterase', *Biochemistry*, 7, pp. 4005-10.
6. Perombelon, M.C.M., and Kelman, A. (1980), 'Ecology of the soft rot Erwinias', *Ann. Rev. Phytopathol.*, 18, pp. 361-87.
7. Pressey, R. (1986), 'Changes in poly-galacturonase isoenzymes and converter in

tomatoes during ripening', *Hort. Science*, 21, pp. 1183-5.

8. Pressey, R., and Avants, J.K. (1973), 'Two forms of polygalacturonase in tomatoes', *Biochem. Biophys. Acta*, 309, pp. 393-69.

9. Tucker, G.A., Robertson, N.G., and Grierson, D. (1980), 'Changes in polygalacturonase isoenzymes during the ripening of normal and mutant tomato fruit', *Eur. J. Biochem.*, 112, pp. 119-24.

19 Proteolytic Enzymes

The term 'protease' refers to all enzymes that hydrolyse peptide bonds. Other names include peptidase and peptide hydrolase. This group of enzymes can be subdivided into exopeptidases and endopeptidases for exo-acting and endo-acting patterns. Endopeptidase is used synonymously with proteinase.

Proteinases are classified into four groups according to the catalytic residue involved in the nucleophilic attack at the carbonyl carbon of the scissile bond: serine (EC. 3.4.21), cysteine (EC. 3.4.22), aspartic (EC. 3.4.23), and metallo-proteinases (EC. 3.4.24).

19.1 CYSTEINE PROTEINASE

This group of enzyme has, in common, the ability to hydrolyse the peptide bonds of proteins and inhibition by sulphydryl reagents. The group of enzymes includes the higher plant enzymes, papain (EC. 3.4.4.10) from papaya, ficin (EC. 3.4.4.12) from fig, and bromelain (EC. 3.4.4.24) from pineapple.

Papain is a single polypeptide that contains 212 amino acids with a calculated MW of 233,500. The sequence consists of high Gly 28, Tyr 19, and Val 18. There are three disulphide bonds: Cys 22-Cys 63, Cys 56-Cys 95, and Cys 153-Cys 200, but only one free thiol group, which is the catalytic Cys 25.

19.1.1 Mechanism of Catalysis

The general catalytic mechanism of cysteine proteinases is represented by two half reactions (Figure 19.1).
1) Acylation: Transfer of the acylating group (specificity side) of the substrate to the reactive Ser of the enzyme.

2) Deacylation: The acyl group in the acyl-enzyme replaced by water to become the leaving group.

The two half-reactions consist of the following reactions.
1) The enzyme binds reversibly with the substrate to form a Michaelis complex.
2) Nucleophilic attack by the Cys 25 thiolate anion on the carbonyl carbon of the scissile bond of the substrate leads to formation of a tetrahedral intermediate.
3) A collapse of the tetrahedral intermediate to an acyl-enzyme is facilitated by protonation of the leaving group by the His 159 imidazolium.
4) Nucleophilic attack by a H_2O molecule assisted by His 159 acting as base catalyst results in deacylation of the acyl-enzyme.

19.1.2 Optimum pH and Temperature

The enzyme is exceptionally stable to high temperatures at neutral pH, but denatured at acidic pH < 4. The optimum pH of papain is 5.5-7.0 and its pI is 9.6. It is also resistant to denaturation in concentrated (8M) urea solution at neutral pH range. Papain retains its activity in various organic media. In fact, the enzyme is extensively investigated for organic synthesis of peptides and other chemical compounds. The high stability of papain toward various extreme conditions and its availability at affordable cost has made it widely used as a meat tenderizer and in chill haze prevention in the brewery industry.

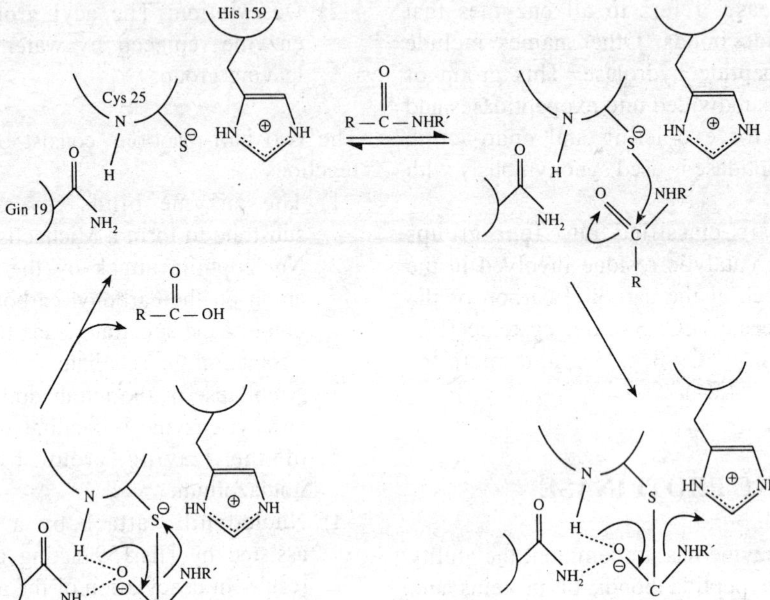

Fig. 19.1 *Reaction mechanism of papain showing acylation and deacylation.*

19.2 ASPARTIC PROTEINASE

The best studied of this group of enzymes is enzyme pepsin. Rennin is the other most useful enzyme commercially because of its wide use in cheese production.

Pepsin is formed by an autocatalytic reaction from pepsinogen, which is found in the stomach mucosa of animals. Pepsin is composed of a single polypeptide chain of 321 amino acids and has a molecular weight of 35,000. Its tertiary structure is stabilized in part by three disulphide bridges and a phosphate linkage. The phosphate group, attached to the hydroxyl group of a seryl residue, can be removed without loss of enzymatic activity.

19.2.1 Mechanism of Catalysis

The free enzyme has two carboxyl groups, one in the protonated form and the other in the ionized form, in the transforming locus of the active site. The enzyme-substrate absorptive complex is formed, followed by a nucleophilic attack of the carboxylate group on the carbonyl group of the peptide bond. This leads to formation of a covalent tetrahedral intermediate. The carbonyl oxygen of the protonated carboxyl group then serves to extract a proton from the hydroxyl group facilitating an electrophilic attack of the carbonyl carbon on the NH group of the peptide bond. The result is the formation of an amino-acyl-enzyme intermediate which then reacts with water to give the products of the reaction (Figure 19.2).

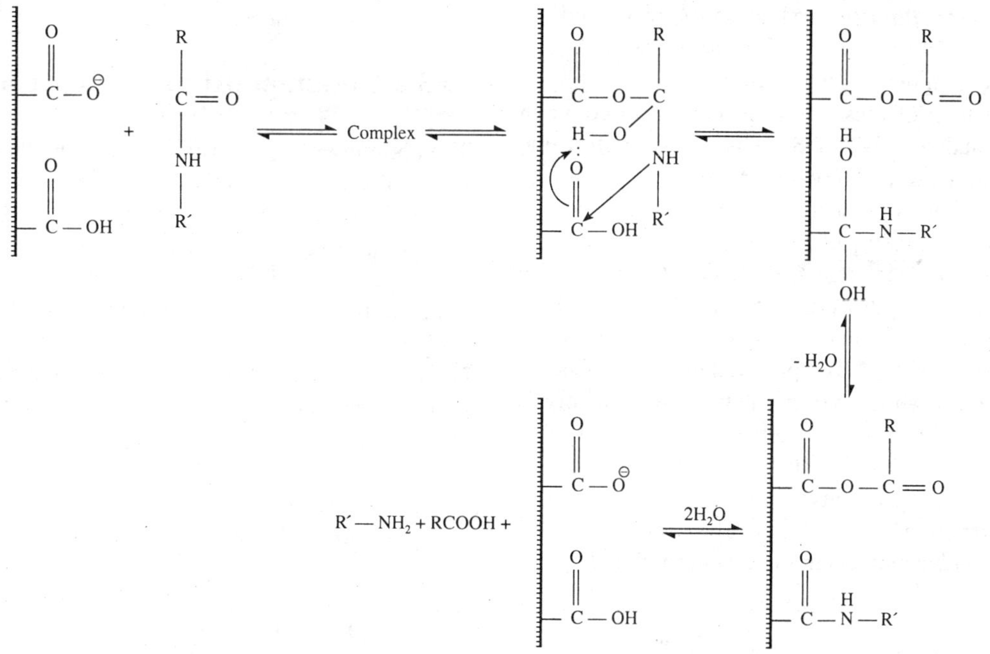

Fig. 19.2 *Proposed mechanism for the action of pepsin.*

19.2.2 Optimum pH and Temperature

The pH-activity profile of pepsin is remarkable because of the very low pH optimum at pH 2 for most substrates. Pepsin is relatively heat stable and can be used at temperatures up to 60°C. Its low pH-activity profile is an advantage in process hygiene.

19.3 SERINE PROTEASES

The serine proteases include the chymotrypsin family, trypsin family and the subtilisin family. Chymotrypsin and trypsin are formed in the digestive tract by autocatalytic reactions in which trypsinogen is hydrolysed to trypsin by the action of trypsin itself. Unlike trypsin and chymotrypsin, which are mammalian enzymes, subtilisin is a group of alkaline serine proteases secreted by species of *Bacillus*. Alcalase is produced by *Bacillus licheniformis* and is the protease that is produced in the largest amounts.

Since all of these enzymes consist in common of a catalytic triad of Ser-His-Asp, therefore the general features of catalysis by this group of enzymes are identical; only the specific groups involved in binding of substrate are different which results in the different specificities. Chymotrypsin catalyses the hydrolysis of peptide bonds in which the peptide carbonyl group is contributed by the aromatic residues like phenylalanine, tyrosine or tryptophan while trypsin catalyses the hydrolysis of peptide bonds in which the peptide carbonyl group is contributed by lysine or arginine. Alcalase has a broad specificity and cleaves many types of peptide bonds, preferentially those with a hydrophobic side chain on the carbonyl side.

19.3.1 Mechanism of Catalysis

Serine proteinases share a number of similar features in reaction mechanism with cysteine proteinases. The general mechanism of both proteinases involves acylation and deacylation with similar basic chemistry. Serine-catalysed reaction consists of the following steps (Figure 19.3).

1) Formation of enzyme-substrate complex (ES) that is non-covalent and reversible.
2) Formation of a covalent tetrahedral intermediate via nucleophilic attack by the reactive serine 221 on the carbonyl carbon of the scissile bond of the substrate. This step is facilitated by general base catalysis by His 64.
3) A collapse of the intermediate to an acyl-enzyme via the protonation of the leaving group (C-terminal segment) of the substrate by His 64.
4) Nucleophilic attack by a H_2O molecule, assisted by general-base involving His 64. This results in the formation of another tetrahedral intermediate.
5) Breakdown of the intermediate via His 64-catalysed protonation of the Ser 220, liberating the acylating side (N-terminal segment) of the substrate as an acid product.

19.3.2 Optimum pH and Temperature

The optimum pH of trypsin is 7.8. At acidic pH, at which the catalytic activity is reduced to zero, trypsin is remarkably stable; it can be stored in the cold in dilute solutions at pH 3.0 for weeks without loss of activity. Trypsin is heat stable and can typically be applied at temperatures 50-60°C.

Alcalase has a broad pH-activity profile with a maximum at pH 8-9. It is quite thermostable, the typical temperature of application is 55-60°C but the enzyme may be applied at temperatures up to 70°C.

In its pH-activity profile, chymotrypsin resembles the subtilisins rather than trypsin.

19.4 METALLOPROTEASE

Thermolysin, which is produced by *Bacillus stearothermophilus*, is the most heat stable of the commercially available proteases. In the presence of Ca^{2+} it is only slowly inactivated, even at temperatures near 80°C. Its pH optimum is slightly alkaline (pH 7-9).

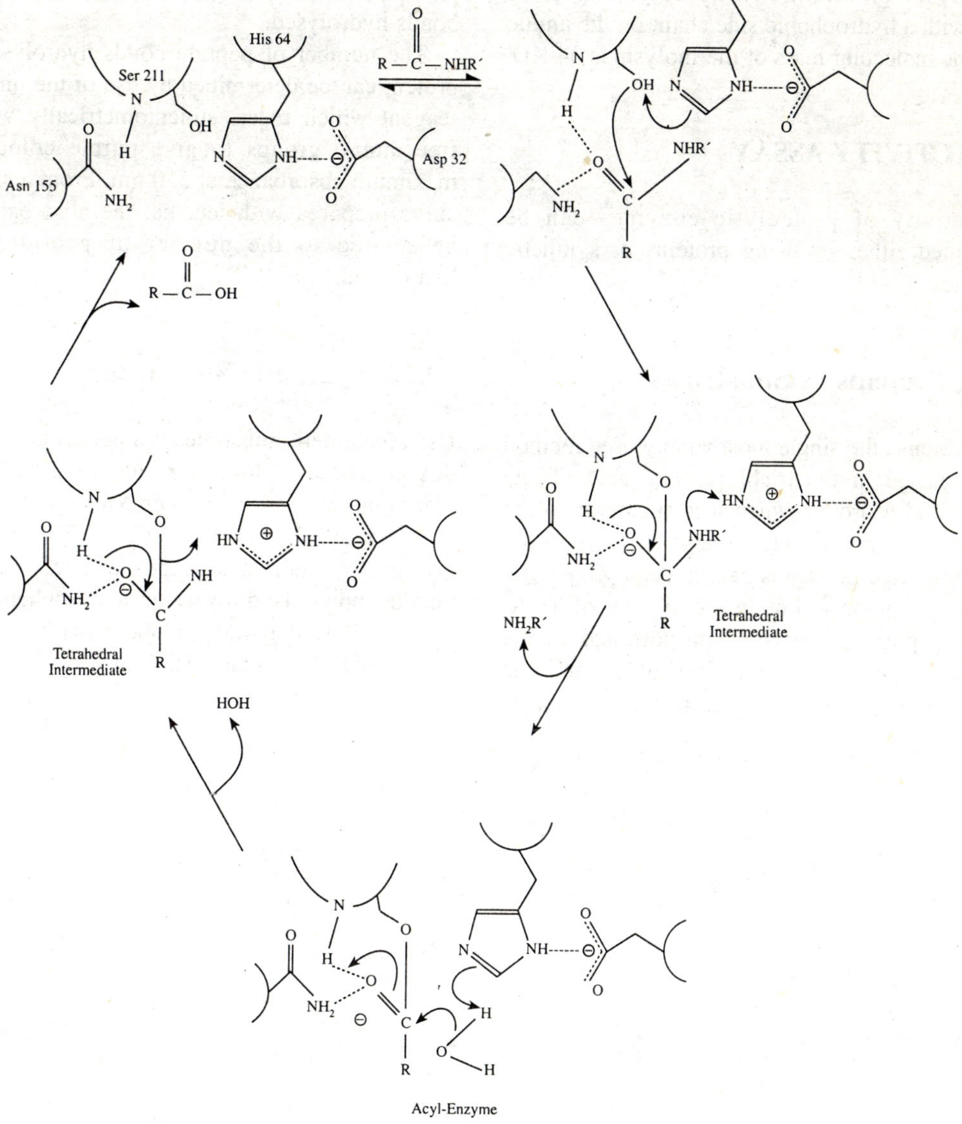

Fig. 19.3 *Reaction mechanism of subtilisin showing acylation and deacylation.*

Like other metalloproteases, thermolysin is inhibited by chelating agents such as EDTA, citrate and phosphate, which is of concern in food uses. With respect to peptide bond specificity, thermolysin preferentially hydrolyses peptide bonds with a hydrophobic side chain on the amino side. The molecular mass of thermolysin is 34 KD.

19.5 ACTIVITY ASSAY

The activity of proteolytic enzymes can be determined either by using proteins or synthetic substrates.

19.5.1 Proteins as Substrates

With proteins, the single most widely used method is the change in the trichloroacetic acid (TCA) solubility of a protein when it is subjected to the action of a proteolytic enzyme. The most commonly used protein is casein. As a proteolytic enzyme acts upon a protein the amount of TCA-soluble peptide produced is proportional to the amount of the enzyme and time of action. The amount of TCA-soluble products formed can be determined by measuring the absorbance of the supernatant liquid at 280 nm or by use of colour reactions involving tyrosine or the peptide bonds in the soluble peptides. The method is fast and precise but does not give the number of peptide bonds hydrolysed.

The number of peptide bonds hydrolysed in a protein can be determined by use of the ninhydrin reagent which reacts stoichiometrically with the free amino groups to give purple colour with maximum absorbance at 570 nm. From a standard curve prepared with leucine, the absorbance can be related to the number of peptide bonds hydrolysed.

19.5.2 Synthetic Substrates

Use of synthetic substrates is a necessity in studies designated to elucidate the specificity and mechanism of action of an enzyme.

Most synthetic substrates for proteolytic enzymes contain a susceptible ester, amide or peptide bond. Hydrolysis of a nitrophenyl ester can be followed at 400 nm (above pH 7) or 340 nm (below pH 7) (equation 1).

$$R-C(=O)-OR' + H_2N-OH \xrightarrow{OH^-} R-C(=O)-NOH \cdot H \xrightarrow{Fe^{3+},\ H^+} \text{Red colour read at 540 nm} \quad (3)$$

$$+ \ R'OH$$

Frequently, hydrolysis of other esters or amides can be followed spectrophotometrically. For example, hydrolysis of α-N-benzoyl-L-arginine ethyl ester (or amide) shows an increase in absorbance at 253 nm due to formation of the carboxylate group (equation 2).

Hydrolysis of an ester at a pH above the pK of the carboxyl group formed results in the liberation of a proton. This can be followed conveniently in a pH stat which automatically adds base to neutralize the proton produced. At pH values either below the pK of the carboxyl group formed or above the pK of amino group (or ammonia) formed, hydrolysis of an amide substrate can also be followed in a pH stat.

The rate of hydrolysis of an ester can also be followed conveniently by the alkaline hydroxylamine-$FeCl_3$ reaction by removing aliquots of the reaction periodically and determining how much ester is left. Hydroxylamine reacts with the remaining ester in alkaline solution to give a hydroxamate which in turn gives a red colour with Fe^{3+} in an acid solution (equation 3).

19.6 APPLICATIONS IN FOOD PROCESSING

The uses of industrial proteases in foods fall in two different categories:

1) processing aids in the manufacture of traditional food products, and
2) production of new food protein ingredients with particular functional properties.

19.6.1 Proteases as Processing Aids

19.6.1.1 Production of cheese

The classical example of the use of a protease is in cheese making, where rennet is added to gel the casein in milk. Rennet, a mixture of chymosin, also called rennin, and pepsin, is obtained from the gastric mucosa of young animals, e.g. calves and lambs. The pepsin to chymosin ratio of different rennet preparations can vary considerably because chymosin is present only in the stomachs of unweaned mammals and is later replaced by pepsin. The content of pure chymosin depends on the age and species of the animal. Only chymosin is able to convert specifically casein from the solution to the gel state. Pepsin and other proteolytic enzymes are much less specific and give rise to a number of degradation products which tend to taste bitter. For this reason, pure chymosin and high-quality rennet are important.

Chymosin is a highly specific endoproteinase. It splits only γ-casein into a glycomacropeptide and para- γ-casein by selectively cleaving the 105-

106 bond between phenylalanine and methionine. The enzyme destroys the protective collodial function of γ-casein. As a result, the surface of the casein micelle becomes more hydrophobic, leading to gel formation.

19.6.1.2 Chill-proofing of beer

Papain has been used in the beer industry for prevention of haze formation during cold storage of the beer, the so-called chill-proofing of beer. The effect of the enzyme is believed to be the proteolytic degradation of ill-defined complexes between proteins and tannins.

19.6.1.3 For meat tenderization

The deliberate action of proteases on muscle tissue to make it softer has been practiced for centuries in the form of cooking meat wrapped in papaya leaves. Papain is still the most important enzyme used for meat tenderizing although a considerable number of enzymes have been promoted in this area. The simplest way to apply the enzyme is to sprinkle it over slices of raw meat.

Fish meat is much less tough than beef. When proteases are applied to fish muscle, the desired result is generally a thorough degradation of the fish muscle into soluble peptides. The traditional autolysis processes leading to the various fish sauces of Southeast Asia exemplify this application.

19.6.1.4 Synthesis of aspartame

α-Aspartame
(Sweet)

β-Aspartame
(Bitter)

The synthesis of aspartame, a low-calorie sweetener, has attracted much attention, especially since its approval by FDA. Aspartame is a dipeptide consisting of L-aspartic acid and the methyl ester of L-phenylalanine. Its sweet taste depends on the L-conformation of the two amino acids, the presence of the methyl ester, and the correct coupling of the amino acids. α-Aspartame is 150-200 times sweeter than sucrose, whereas β-aspartame has a bitter taste.

L - Z - Asp   D,L - Phe - OCH$_3$

Thermolysin

L - Z - Asp — L - Phe - OCH$_3$ D - Phe - OCH$_3$

Hydrogenation   Recemization

L - Asp — L - Phe - OCH$_3$ D,L - Phe - OCH$_3$

α-Aspartame
where Z = benzyloxycarbonyl

Aspartame is currently produced by a chemical process in which the required stereospecificity adds to the production costs. These disadvantages can be overcome by an enzymatic approach. The enzyme thermolysin, a neutral metalloprotease from *Bacillus thermoproteolyticus*, is particularly suited for the synthesis of aspartame. Under appropriate conditions, it catalyses the condensation of N-protected L-aspartic acid and D,L-phenylalanine methyl-ester. Reactions occur exclusively at the α-carboxylic group of aspartic acid and with the L-isomer of phenylalanine methyl ester.

19.6.2 New Functional Protein Ingredients

An entirely different purpose of proteases is pursued in the production of new products where quality is the goal. Unconventional protein sources can be modified by protein hydrolysis to improve functional properties with respect to a particular use in food systems. Thus, an increased solubility and a concomitant improvement of the emulsifying or foaming properties can be achieved with many different food proteins.

19.6.2.1 Protein hydrolysates

Whenever protein hydrolysis is considered, one of the first questions raised is that of bitterness. It is an accepted fact that proteolysis can give rise to a bitter taste which is associated with hydrophobic peptides. Methods of bitterness control are of three types: masking, removal and prevention.

19.6.2.1a Soy protein hydrolysate

Although widely used in the food industry, the functional properties of soy protein are not always those desired. Enzymatic modification can be used to produce products of improved functionality compared with the untreated protein. There are two classes of soy protein hydrolysate, the highly functional hydrolysate and the highly soluble

hydrolysate. For the first class, it is suggested that the low rate of inclusion of this product reduces the impact of any off or bitter flavours. The highly soluble hydrolysates, on the other hand, are used at relatively high concentration and therefore strict control of bitterness is necessary.

Highly functional hydrolysates may be prepared from soy isolate or concentrate. The hydrolysis step is performed at 50°C, pH 8.0 in a pH stat and with an enzyme dose of 200 grams alcalase per kilogram of protein. Hydrolysis to a degree of three results in better emulsification properties, iso-electric solubility values and whipping expansion of the soy hydrolysate.

Functional Properties of soy hydrolysate

Soy isolate		
	Before Hydrolysis	*After Hydrolysis**
Iso-electric solubility	5%	42 %
Emulsification capacity	100 ml /g	280 ml /g
Whipping expansion	20%	200 %

* Degree of hydrolysis 3

Protein hydrolysate with a much higher degree of solubility require more extensive hydrolysis. A degree of hydrolysis of 10 has been found to give high solubility, good protein yield and low bitterness when soy protein is treated with alcalase. The resulting hydrolysate has been used to produce protein-fortified soft drinks and in the production of special dietetic feeds for the hospital sector, in particular for patients suffering from cancers which affects their eating patterns. If soy concentrate is the starting material, then the resulting hydrolysate may be incorporated into calf milk replacers.

19.6.2.1b Milk and whey protein hydrolysate

Hydrolysis of milk proteins presents problems because of the non-specific clotting action of the protease and the tendency for casein hydrolysates to be very bitter. Hydrolysis of casein is, however, possible and casein hydrolysates can be readily prepared by the use of proteolytic enzymes such as alcalase and neutrase.

Casein hydrolysates are largely prepared for incorporation into dietetic products. In cystic fibrosis, enzyme secretion by the pancreas is impaired and tryptic and peptic hydrolysates of casein are used in the treatment of this illness.

19.6.2.1c Fish protein hydrolysate

Fish protein is available for modification in two forms, as waste from fish processing and as fish not currently utilized in human food but suitable for fish meal. In both cases, proteolytic enzymes can be used to produce soluble fish hydrolysates of high nutritional value. As the endogenous levels of proteolytic enzymes vary with fish type and season, external proteases are used to control the hydrolysis reaction.

Many different proteases have been used for hydrolysis, including both plant and microbial enzymes. Thermostable enzymes are preferred as a mean of microbial spoilage, giving processing temperatures in the range 55-65°C. Highest yields are obtained when the hydrolysis is carried out at alkaline pH. Under these conditions significant solubilization of protein occurs.

The resulting protein hydrolysate has been evaluated as a component of calf milk replacer where it has been found to replace successfully 100 per cent of the milk proteins for lambs, and 50 per cent of the milk protein in calf feeding.

SUMMARY

The term protease refers to all enzymes that hydrolyse peptide bonds. They are classified into four groups according to the catalytic residue involved in the nucleophilic attack at the carbonyl carbon of the scissile bond: serine, cysteine, aspartic and metalloproteinases.

The group of enzymes includes the higher plant enzymes, papain (EC. 3.4.4.10) from papaya, ficin (EC. 3.4.4.12) from fig, and bromelain (EC. 3.4.4.24) from pineapple.

The general catalytic mechanism of cysteine proteinases is represented by two half reactions: (1) acylation and (2) deacylation. The enzyme is exceptionally stable to high temperatures at neutral pH, but denatured at acidic pH < 4.

The best studied of aspartic enzymes is pepsin. Rennin is the other most useful enzyme commercially because of its wide use in cheese production. The pH-activity profile of pepsin is remarkable because of the very low pH optimum at pH 2.0 for most substrates. Pepsin is relatively heat stable and can be used at temperatures up to 60°C. Its low pH-activity profile is an advantage in process hygiene.

The serine proteases include the chymotrypsin family, trypsin family, and the subtilisin family. Since all of these enzymes consist in common of a catalytic triad of Ser-His-Asp, therefore the general features of catalysis by this group of enzymes are identical; only the specific groups involved in binding of substrate are different which results in different specificities. Serine proteinases share a number of similar features in reaction mechanism with cysteine proteinases. The general mechanism of both proteinases involves acylation and deacylation with similar basic chemistry. The pH optimum of trypsin is 7.8. At acid pH, at which the catalytic activity is reduced to zero, trypsin is remarkably stable; it can be stored in the cold in dilute solutions at pH 3.0 for weeks without loss of activity. Trypsin is heat stable and typically can be applied at temperatures of 50-60°C.

Thermolysin, a metalloprotein, which is produced by *Bacillus stearothermophilus*, is the most heat stable of the commercially available proteases. Its pH optimum is slightly alkaline (pH 7-9). With respect to peptide bond specificity, thermolysin preferentially hydrolyses peptide bonds with a hydrophobic side chain on the amino side.

The activity of proteolytic enzymes can be determined either by using proteins or synthetic substrates.

Applications of proteases in food processing include production of cheese, chill-proofing of beer, meat tenderization, synthesis of aspartame, and production of protein hydrolysates.

RECOMMENDATIONS

1. Beddows, C.G., and Ardeshir, A.G. (1979), 'The production of soluble fish protein solution for use in fish sauce manufacture, I, The use of added enzymes', *J. Food Technol.*, 14, pp. 603-12.
2. Cunningham (1965), 'The structure and mechanism of action of proteolytic enzymes', *Comprehensive Biochemistry* (M. Florkin and E.H. Stotz, eds.), 16, p. 85, Elsevier Publishing Co., Amsterdam.
3. Dayhoff, M.O. (1969), 'Atlas of protein sequence and structure', 4, p. 49.
4. Moll, M. (1987), 'Colloidal stability of beer', *Brewing Science* (J.R.A. Pollock, ed.), 3, pp. 1-327, Academic Press, London.
5. Reichelt, J.R. (1983), 'Baking, Industrial Enzymology', *The Application of Enzymes in Industry* (T. Godfrey and J. Reichelt, eds.), pp. 210-20, MacMillan, London.
6. Schwimmer, S. (1981), *Source Book of Food Enzymology*, AVI, Westport, Connecticut.
7. Ward, O.P. (1983), *Proteinases Microbial Enzymes and Biotechnology* (W.M. Fogarty, ed.), pp. 251-317, Elsevier Applied Science, London.

Index